Strategy for sustainable business

Strategy for sustainable business

Environmental opportunity and strategic choice

Liz Crosbie
Ken Knight

McGRAW-HILL BOOK COMPANY

London · New York · St Louis · San Francisco · Auckland
Bogotá · Caracas · Lisbon · Madrid · Mexico · Milan
Montreal · New Delhi · Panama · Paris · San Juan
São Paulo · Singapore · Sydney · Tokyo · Toronto

Published by McGraw-Hill Book Company Europe
Shoppenhangers Road, Maidenhead, Berkshire, SL6 2QL, England
Telephone 01628 23432
Fax 01628 770224

British Library Cataloguing in Publication Data
Crosbie, Liz
Strategy for Sustainable Business. –
(Henley Management Series)
I. Title II. Knight, Kenneth III. Series
658.408

ISBN 0-07-709133-7

Library of Congress Cataloging-in-Publication Data
Crosbie, Liz,
Strategy for sustainable business: environmental opportunity and strategic choice / Liz Crosbie, Ken Knight.
p. cm.
ISBN 0-07-709133-7 (hardback)
1. Social responsibility of business. 2. Environmental auditing. 3. Strategic planning. 4. Sustainable development. I. Knight, Kenneth. II. Title.
HD60.C76
658.4′08–dc20 95-23580
CIP

Typeset by BookEns Ltd, Royston, Herts.
and printed and bound by Biddles Ltd, Guildford and King's Lynn

Contents

PART TWO FOUNDATIONS OF ENVIRONMENTAL STRATEGY

PART THREE HANDBOOK OF FUNCTIONAL ENVIRONMENTAL STRATEGIES

Preface

This book is about the environment and business strategy – how can environmental management be made more strategic, so as to become an integral part of the mission, the objectives and the planning processes of the company? It is also about sustainability – how can 'sustainable development' be made a reality at the level of the individual business? Are these two separate questions? We believe that they turn out to be one and the same.

Environmental management is a new discipline, and to date the bulk of progress has been in expanding the range of control techniques on offer. The more techniques proliferate, the greater the need for an overriding framework. Under public pressure to improve their environmental performance, companies have had little time to create a structured decision-making system based on a consciously chosen strategy. When it comes to their relations with the natural environment, few businesses as yet formalize their strategic choices or monitor the opportunity cost of alternative options.

What is becoming essential now is more than a competitive strategy for the business, and more than an environmental strategy to cope reactively with outside pressures. The need is for an integral 'business environment strategy', concerned with the future of the business as a whole in the context of an 'environmental revolution' that threatens – or promises – to be one of the major determinants of business development as well as corporate survival in the coming decade.

In this book we explore the scope of environmental management to show what potential developments are possible for companies who have reached, or are moving towards, a level of strategic maturity in dealing with the environment.

At present it is rare for environmental strategy to be integrated into, and to inform all aspects of, the business's commercial strategy. Many companies still routinely separate environmental objectives from business objectives.

And if integration is rare, the pursuit of sustainability is a route even fewer now follow. The emergence of protocols for environmental management such as British Standard 7750 and the European Eco-Management and Audit Scheme (EMAS), have led many to confuse the pursuit of sustainability with 'continuous improvement'. But without the adoption of full environmental sustainability as an overriding operating goal by the individual firm it is difficult to see how the objective of a sustainable market economy can be achieved.

The issues of strategy and of sustainability arise at more than one level. Strategic decisions at the corporate level may not find their way to full implementation by business units subjected to conflicting priorities, and environmental 'strategy in action' requires the adoption of compatible strategies by the different functions and specialisms of management.

Currently there is insufficient available material describing how sustainable development will work at the level of the individual business, let alone the level of marketing or research and development (R and D) or operational management within the individual firm. Much of the detail of everyday sustainable working has yet to be described or developed, and doing this work in practice is how understanding will grow. As a contribution to this new discipline of sustainable business operation, this book raises some of the fundamental questions that need to be answered, but also seeks to set out many of the 'nitty-gritty' operational implications for the benefit of practitioners. To reflect this dual role, conceptual and practical, the book is divided into sections. The early parts are particularly intended to stimulate radical reassessment of current thinking about business and the environment, while the later sections are designed to be of immediate use to directors and managers in developing their companies' environmental position and practice. There is a summary of the core argument at the beginning of each chapter, to enable the reader to use the book in the most effective way.

The book has four parts. Part One provides the context – to clarify why a new corporate goal of sustainable operation is both necessary and urgent. Part Two lays the foundations of a corporate environmental strategy – the strategic options, the strategy development process and the financial policies and systems needed to underwrite the strategy. Part Three offers a practical handbook of functional strategies of ecological management – managing different aspects of the business in a way that is both compatible with ecological systems and financial reality. Finally, Part Four provides the basis of an implementation strategy.

Liz Crosbie and Ken Knight

Acknowledgements

We are very grateful for the generous help, in time and information, which we have received from many people in and outside business, and would like to record our special thanks to Dr Harry Adam, John Carew, Ian Christie, Professor Roland Clift, Diana Montgomery, Dave Sharman, Hugh Smith, Roger Twiney, Dr David Wheeler, Dr Bob Whelan, Dr Jonathan Williams, and the AMED Sustainable Development network.

We would also like to thank Fritjof Capra, Paul Chapman Publishing Ltd., *The Columbia Review of World Business*, Alan Dale, Victor Gollancz Ltd., Greenleaf Publications, *New Scientist*, Prentice Hall International (UK) Ltd. and Shell International for permission to reprint material from their publications.

We are particularly grateful to Nickie Fonda for reading the manuscript and advising on how it could be improved. Above all we must thank our long suffering families for their support throughout this project.

Any mistakes and oversights remain ours alone. In this fast moving field there will be, no doubt, many issues and examples of good practice omitted. We hope you will enjoy using this book and we look forward to recording your own progress towards sustainable business in the future.

Liz Crosbie and Ken Knight

PART ONE
Becoming sustainable – the challenge to business

In Part One we explore the context for business as a whole, reviewing how wider society views its activities and is dictating changed environmental priorities. We also look at the context of the individual business, the operational unit trying to make sense of revised external expectations.

Society's demands, as well as the scope of business, are rapidly changing. World trade and business dealings are being internationalized. Global awareness of the environmental impacts of commercial activity is increasing. We write, therefore, from an international perspective, while selecting examples from those countries where most innovation is taking place.

These emerging realities are reflected in our first three chapters.

Chapter 1 looks at the new *strategic* importance of the environment for today's business.

Chapter 2 documents society's growing demand for business to become environmentally sustainable and what this implies.

Chapter 3 argues that to meet this demand will require a new approach to management.

1 A strategic approach to the environment

1. The environmental challenge is much more than a question of legal compliance, it is a major strategic issue for business.
2. Companies are facing real strategic choices, affecting their competitive future and even their survival.
3. Human impacts are damaging our natural life support systems to an extent that endangers our future on the planet.
4. The threats to the natural environment are bringing about critical changes in many aspects of the business environment, which call for a new strategic response.

1.1 Whose health and safety?

'Multi-billion-dollar company responds to threat of global collapse by extending Health and Safety department'.

We have not actually read this headline in the business press, but we could have done – many times over. The predominant response of business to the growing deterioration of our natural environment has been simply to strengthen its internal compliance functions. Please do not misunderstand; we think that prevention of industrial accidents and in-company control of 'substances hazardous to health' are vital functions. We also understand only too well the business logic that delegates environmental protection to the Safety Officer.

'We are not a charity,' runs the argument. 'Our duty, to our shareholders, to our employees and to society, is to succeed as a business, while being responsible corporate citizens and obeying the law. Protecting the environment is the government's responsibility, and they do it by passing laws for us to abide by. To ensure that we comply with those laws and don't harm the health of our neighbours is precisely what we have a Health and Safety department for, and they're very good at it. So we'll turn it into the Health, Safety and Environment department, or even, to show how committed we are, the Environment, Health and Safety department!'

The logic of the argument is impeccable, given the premise that environmental protection has nothing to do with what the business is about, and that the real business is that of making money by competing in the marketplace. But it is that premise that needs to be challenged. Those who accept as an article of faith that companies are only there to make money are wrong. In the most basic terms, most companies make money because they satisfy the needs of the society which provides them with a market. (There are, of course, niche markets for satisfying antisocial needs, but they should emphatically not be taken as the norm.)

Today, society in the developed world is becoming aware that what it needs most is to preserve a clean, life-supporting environment. And even in developing countries, in spite of often appalling living standards, this need is beginning to come to the fore. Being able to breathe is as important as being able to eat.

Gearing and steering the company to satisfy broad social goals expressed through evolving market needs is *business strategy*, and it is the job of senior management and the board of directors. Of course, legal compliance is also important and watchdogs are needed. But leaving the environment entirely to the watchdogs means abdicating the strategic responsibility for a major new business issue. The environmental challenge to business is not only about the health and safety of employees and the local community, but also about the health today and safety tomorrow of the business itself, not to mention the market economy as a whole, society at large and our planet.

1.2 What do you mean – strategic?

Business gains its customers, and its legitimacy as an activity, from meeting society's needs. As a result of our newly acquired knowledge of environmental damage to the global systems that support life, society has begun to express overtly its need for a clean, wholesome and stable environment. The potential business opportunities inherent in meeting this demand are vast and must be considered alongside the more obvious constraints on operating practice that such changes will bring. It is this combination of opportunity *and* pressure for major change that makes the environment a strategic issue *par excellence*.

Strategy is notoriously about making choices, about setting goals and choosing the paths for reaching them. This book seeks to highlight some of the choices that are involved in a strategic response to the environmental challenge. This emphasis on choice is deliberate. There is much in print telling managers how to be 'green'; from these texts it is possible to assemble a coherent set of prescriptions for environmental management, which make good environmental sense and will command wide-ranging support. Managers read these prescriptions, and agree to them, for as long as they are focusing their attention primarily on the environmental aspect of their job. 'Yes,' they will say, 'this is obviously what we *should do* if we can.' And then they will go back to doing what they *have to do* within the constraints of the way their business is structured.

What this ready acceptance of 'real world' constraints overlooks is that the way the business is structured is the way the participants have chosen that it should be structured, and that those choices can sometimes be changed. As and when companies decide that environmental 'sustainability' is an overriding, non-negotiable objective, they will almost inevitably be forced to look again at how their business is organized. Likewise, almost the whole of business will have to be organized differently to meet this objective.

So in focusing this book on strategic choice we are particularly concerned with the different ways of developing a successful business once the need for environmental sustainability has been *accepted* – not forgetting, of course, that there can also be a strategic choice to ignore the environment and go for short-term advantage, with all the social and economic consequences of so doing.

Or there can be an apparent failure to make any strategic choices at all, which is a way of glossing over or concealing the fact that the choices which are actually being made are neither reputable, ethical nor sensible in the long run. Such choices as ignoring the environment, fighting against attempts to improve it, making money from destruction of resources which are essential to survival, supporting strong-arm tactics against defenders of natural resources, are easiest to make by pretending to have no strategy. As soon as the fact that these routes are actual strategic options, and that there are others, is made explicit and laid squarely on the table, such a response becomes much more difficult for a reputable company to maintain.

But in discussing the environment with managers, we have been surprised how few regard it as a strategic issue at all. To date, many are responding to increased consumer and regulatory pressure for change with non-strategic responses: bolting on Environment to the Health and Safety remit as a legal necessity, or engaging in what is politely known as 'greenspeak', i.e. talking environment while doing very little. In the former case the environment is seen as another regulatory challenge, rather than a shift in the overall operating context. In the latter the demands of outside stakeholders for information have generated a new extension of the public relations department, with increased presentation and information flows but very few changes to the working practices of the company.

We believe that limited responses such as these are part of a transition process, as individuals and companies try unsuccessfully to deny the full implications of the changes that need to take place. A great shift in thinking is required, and it is no surprise that for many the conceptual leap is very painful.

1.3 A new environment for business strategy

During the training of most people associated with corporate strategy, 'environment' originally meant something very different. Pick up any of the many management textbooks on strategy and, under the heading of 'environment', you are likely to be bombarded with information on competitors' business plans and products, economic indicators, legal changes and social factors, and very little to do with the basic support systems on which we, as a species, are dependent for our survival.

The new awareness of how close human beings have come to exceeding the overall carrying capacity of the planet's systems was occasioned by the global warming and ozone depletion debates, and these did not reach the wider public until well into the 1980s. This knowledge of human impacts on the climatic, carbon and solar energy cycles changed the operating context for all human beings. No-one living on the planet, however remote, can now escape the effects of our species 'experiment with life itself', to quote Margaret Thatcher, and there are few who can still claim real ignorance about the risks. The 'business environment' must today include the living, natural environment as a vital context in which decisions are reached.

Traditionally the business's operating environment was seen to consist of a series of subsectors. These sectors build up, layer upon layer, a model of the business world that its participants were able to control or indirectly influence:

- Markets: local and global, consumer and industrial, and competition within them
- Supply factors, affecting materials, components and products
- Science and technology
- Demographic trends
- Economics and finance
- International relations and trade
- Politics and legislation
- Social trends and values.

Each of these 'sub-environments', and their implications for business strategy, are now affected and changed by what is happening to the natural environment.

1.3.1 The impacts of human activity

Table 1.1 attempts to prioritize the most important natural issues over the foreseeable future from a business perspective. (This outline is based on available knowledge of the limits on the 'sinks' available for waste materials or energy flows in the global system.[1])

Table 1.1 Environmental threats

Global warming and energy use

One of the greatest known dangers to the planet's systems is human activity changing the mix of gases in the atmosphere by the amount of fossil fuel being burnt and compounds manufactured. Certain gases restrict the earth's ability to reflect solar energy, thus trapping heat; build-up of this is changing the earth's climate. Business is one of the major contributors of greenhouse gases; changes to the cost structure of energy are a likely response to limit overall use, and this will affect business more than other producers of environmental impacts. Finding ways to reduce energy demand and controlling the users of energy internally is a significant challenge.

Toxic substances and business operations

There are now so many toxic substances released into natural systems that science does not have sufficient knowledge to understand what the ramifications are for our life-support mechanisms, or the dangers of these chemical cocktails to human health. Many of these substances are not broken down by natural processes and continue accumulating indefinitely in the eco-system. Fear about the ability of widely used chemicals and radioactive waste to migrate through food chains and concentrate, causing poisons, has led to legislative control of handling procedures, waste disposal requirements and in some cases outright prohibition. Learning to live with the by-products of our new technologies has been difficult, and as industrial processes become more complex the expectations of wider society to be informed of the risks and for these risks to be controlled will increase.

Ozone destruction and CFCs

Just as many chemicals in use today are thought to be harmless, the chlorofluorocarbons (CFCs) were once wonder molecules. Their inherent stability, so useful on the ground, has been the undoing of the stratospheric ozone layer to which they ascend, decomposing to release chlorine, which attacks the ozone shielding the planet against damaging ultra-violet radiation. Alternatives have been made available for CFCs, but many of the early substitutes are simply contributing to the greenhouse effect and the net environmental gain overall is minimal.

Acid rain and transport

The water cycle on which we all depend is now out of balance in much of the western world; lakes and watercourses are being acidified and forests attacked by airborne pollutants, generated by energy production, the chimneys of large-scale industry, and transport systems. Pressure is now on for individual organizations to produce localized solutions to reduce output of acidifying gases, but much of this good work is being undermined by the growing numbers of internal combustion engines producing ever greater amounts of acid-generating contaminates, as the fleets of private and company cars for personal transportation and lorries for business freight expand, pouring out greater and greater amounts of pollution.

Marine pollution and health

Just as we have to manage the sky as an environmental sink, the sea cannot continue to be an endless storage system for our waste. The ability of human beings to simultaneously pollute and plunder the sea of its resources is under international scrutiny. The UN Food and Agriculture Organization (FAO) now estimates that all 17 of the world's major fishing areas have either reached or exceeded their natural limits, and that 9 are in serious decline.[2]

Reduced biodiversity, loss of species and habitats

Loss of the world's major forests is a disaster from a climatic viewpoint not only because trees, which are the lungs of the world, giving us the precious oxygen we need to survive, absorb the greenhouse gas CO_2; but also in view of the plant and animal life that they support that is being lost. 'There may be anything from 10 million to 100 million species of life on earth. Only 1.4 million have been named and classified.[3] Depletion is now running at such a rate that estimates range from 50 to 250 species being lost every day.[4]

Many argue that we have a duty to protect these other life forms for their own sake, rather than sacrifice them to our insatiable demands for ever greater consumption. But even in purely selfish terms each species that is lost may well represent a medical or food base resource and we will now never know what riches have already been lost to us, quite apart from the impact on the quality of life of other species of humanity's plunder. What we must do is to stop the remaining destruction of forests, wetlands and other habitats before many more species lose the capacity to survive, as supporting microclimates and soil structures are also destroyed.

Forests are being lost due to logging and pollution, wetlands due to drainage and salination, bogs due to peat cutting and drainage. Plants and other species are being lost due to destruction of habitats, pollution and global warming changes to climate.

Without biodiversity, the security of food supply is an illusion. The human population is highly dependent upon a very small number of food plants. The genetic inheritance of wild plants is essential, as without this we are unlikely to be able to protect that food base against disease and viruses by interbreeding. Our human security, therefore, is threatened by the loss of habitats and of other species. And in any case, we firmly believe that these other creatures have a right to exist, irrespective of human beings.

1.4 The business environment meets the natural environment

Reviewing the 'business environment' in the light of these developments in natural systems reveals radical changes within its traditional subsectors. Table 1.2 reviews how environmental change may have impacts on societies globally and on organizations trying to function within turbulence of this magnitude.

Table 1.2 The impact of environmental change

Business environment subsector	*Environmental change and business-related impact*
Markets and competition	– New global market for environmental protection – technologies, products, services, software, know-how. – Demand for less environmentally damaging products, and producers, from consumers and institutional buyers, resulting in competitive advantage gained through environmental performance. – Environmental deterioration diverts a greater percentage of income to clean-up. – Companies wish to purchase from suppliers who can underpin their own environmental performance and provide components and supplies to meet agreed environmental criteria.
Supply of raw materials, components, and products	– Increased cost of some raw materials to reflect scarcity, higher costs of transport as local production capacity exceeds supply, increasing costs of clean-up of extraction and processing reflects concern about damage to environment. – Increased demand for materials obtained by sustainable management practices, like forestry, crop cultivation and animal husbandry, and pressure group action against continued use of unsustainably produced materials. – Need to increase performance of sustainable materials to replicate in use the performance criteria of non-

renewable or unsustainable alternatives; creating opportunities for bio-engineering, plant and animal breeding.

Science and technology

- Increasing awareness of ecological systems and how existing technologies impact on them. Discovery of new dangers threatens existing businesses, such as those producing dioxins.[5]
- Development of new cleaner technologies. Increased technology transfer to ensure that 'best available technology' is made available globally as soon as possible.
- Investment demands increase to meet these new challenges; there is a danger that levels of investment may outstrip the productive capacity to create adequate margins to fund that investment.

Demographic trends

- Increase in global population: the world's population has doubled since 1950; in 1993 87 million (roughly equivalent to the population of Mexico) were added to the global population.
- Increases in population are outstripping increases in productivity in much of the world.
- In many global markets there is insufficient realizable demand because of low incomes. Twenty per cent of the world's population shares only 1.4 per cent of global wealth.[6]
- Increased population offers unlimited market potential for companies producing sustainable technology at price levels that are affordable in developing countries.
- Need to reduce material consumption in developed nations creates opportunities to increase 'non-material' living standards.

Economics and finance

- Global governmental commitment to the principle of sustainable development leads to changes in many economic structures, through the use of economic instruments, fiscal incentives and environmental taxes to identify and protect ecological capital.[7] (See Chapter 5.)
- Environmental taxation and increasing clean-up requirements increase cost of doing business, depressing margins in companies with high environmental impacts. Increased costs of waste disposal, energy and water use for all organizations.
- Need for major investment in moving to sustainable business practices may involve acquisition, disposal

	and restructuring of business portfolios. – Increased provision for environmental liability (e.g. land contamination). Cost of insurance will increase, for some industries and in some countries insurance will no longer be available.
International relations and trade	– Subsidized environmental technology transfer to developing countries. – Growth of political instability as countries suffer from environmental degradation and resulting impoverishment, leading to threats of economic migration, trade difficulties and competition, including warfare for scarce environmental resources, such as water in the Middle East. – Threats to international operations and investments from popular unrest and disruption to market development, higher taxation levels to cover aid relief, boundary protection and higher insurance costs from environment-related damage such as storms, hurricanes, coastal erosion and the climate changes due to global warming.
Politics and legislation	– Growth of environmental legislative programmes in most of the OECD countries. Restrictions on the transfer of environmental wastes and dirty processes between OECD countries and emergent economies. – Information disclosure requirements and increased information available to stakeholders, making company decisions and actions more transparent.
Social trends and values	– Growing environmental consciousness is changing attitudes and values. Shifts in consumer purchasing noted from the mid-1980s in many markets, with changes in institutional and business standards following. – Changing attitudes to wider community due to media coverage of global issues and different cultures. Growth of global communications systems allowing transfer of knowledge between communities of corporate activities and reputation.

This limited review of the traditional subsectors of business situation analysis demonstrates the wide-ranging nature of the potential changes brought about by environmental issues permeating wider society. Corporate awareness of environmental issues will develop in line with overall knowledge of environmental impacts. As the global significance of seemingly ordinary behaviours becomes clearer, the pressure on organizations to change those behaviours will increase.

These macro-environmental issues facing business underpin the fundamental challenge to become more *sustainable*, and contrast markedly with the current level of awareness of the issues, as expressed by most companies. An example of the latter is provided by a recent ENDS Special Report,[8] based on a survey of environmental managers in the UK, which outlines their perceptions of the major pressures on companies to appoint environmental managers, as follows in Table 1.3.

Table 1.3 Perceived environmental pressures

Legislation
Costs
Pressure from the financial sector
Customer reactions – pressure from the general public – from companies concerned about their suppliers' environmental performance – from public bodies
Community relations

While this list certainly overlaps the issues which we have outlined and accurately reflects the present level of the environmental debate, it does not look very far ahead in any strategic sense. Above all, it does not indicate that the challenge of 'sustainable business' has been taken on board by many. Understanding the long-term objective of sustainable business operations is vital to designing the transition to be undertaken by both society at large and the individual business.

1.5 Sustainability is more than improvement

In environmental management circles the concept of 'continuous improvement' has been taken over from the Total Quality Movement (TQM) of the 1980s and incorporated into Environmental Management Systems (EMS) (as well as the British Standard BS 7750). While some environmental leaders are turning the concept into sequences of ambitious year-on-year targets, other companies are pleased to discover that it need not be a very onerous requirement, as any improvement, however small, can theoretically qualify.

Current business literature about sustainable development seems to have got stuck at the level of improvement. Implicitly, it seems to be assumed that if businesses continue to improve their environmental performan[illegible] same way as the many inspiring examples quoted, sustainability [illegible] reached and all will be well. However, in view of the enormity of [illegible] problems we have outlined in this and the next chapter, such optimism must, to say the least, be open to doubt.

While there is no ready answer to the question at what point any particular business will have become sustainable, both ecologically and economically (both are essential for the business to survive), we do know that it needs to be asked, urgently, and that at present hardly anyone is asking it. A critical strategic task for managers moving their companies into the twenty-first century is to find answers to this question – to develop the blueprints for a business which can be financially viable while achieving a condition of harmonious coexistence with the living natural environment.

Until a serious effort is made to create such plans, no-one knows whether sustainable business is any more than a 'good idea' (as Gandhi said of Western civilization) and whether our economic system has a future.

References and further reading

1. Alexander King and Bertrand Schneider, *The First Global Revolution: A Report by the Council of the Club of Rome*, Simon & Schuster, 1991. See Donella Meadows, Dennis Meadows and Jorgen Randers, *Beyond the Limits: Global Collapse or a Sustainable Future*, Earthscan, London, 1992, for a fuller discussion of environmental sinks.
2. Lester Brown *et al*, *State of the World 1994*, Earthscan, 1994.
3. *Beyond the Limits: Global Collapse or a Sustainable Future*, op. cit., page 64.
4. P. Hawken, *The Ecology of Commerce: A Declaration of Sustainability*, HarperCollins, 1993.
5. For example, the US Environmental Protection Agency Report on Dioxin has been widely reported as recommending stricter controls on industries generating dioxin carrying air emissions and water effluent. ENDs Report 236, September 1994.
6. Lester Brown *et al*, *State of the World 1994,* Earthscan, 1994.
7. Ernest Von Weizsäcker and Jochen Jesinghaus, *Ecological Tax Reform*, Zed Books, 1992.
8. ENDS Special Report, *Environmental Managers in Business*, published by Environmental Data Services, November 1993.

ability –
ng goal

1. The strategic challenge to business is how to become environmentally sustainable.
2. 'Sustainable development' has become the new international watchword.
3. It is now recognized that the carrying capacity of the planet imposes real limits to human economic growth.
4. Our present economic system is already unsustainable, and we are in danger of multiplying the existing impacts five- and tenfold in trying to bring the developing world and future population up to present OECD standards.
5. The need for change has been accepted by political and business leaders who are beginning to exert real pressure to make it happen.
6. They have started looking at ways of making businesses pay for their use of hitherto free environmental resources.
7. The criterion of sustainability is the return to a cyclical rather than a linear pattern of production.
8. This does not, however, tell the individual business what to do. The actual path to sustainability has to be discovered or invented by each company.
9. Becoming sustainable requires more than continuous improvement. The actual end state of sustainable operation needs to be determined and planned for.
10. What are the criteria for determining whether a company is following a strategy of sustainable development?
11. Can it be sustainable to base business success on creating wants rather than satisfying needs?
12. Understanding the changes in the global operating environment will be essential.

The present rate of global consumption outweighs the laying down of global assets.

'Annual income twenty pounds, annual expenditure nineteen and six, result happiness. Annual income twenty pounds, annual expenditure twenty pounds ought and six, result misery'.

(*David Copperfield*, Charles Dickens.)

2.1 Targeting business sustainability

We believe that environmental sustainability is one of the most vital issues for any organization, as it is primarily about survival – both of the individual business and the species as a whole. The task of business leaders is to ensure that the organizations under their control can play their part in the move towards a sustainable-yield, global economy.

Sustainable business strategy means taking the goal of sustainability – living and working in such a way that human society will be possible for generations to come – and translating that into the changes required of an individual organization – changes which maintain the organization's capacity for producing human benefits, including the profitability needed for survival, while optimizing the environmental balance of its operations.

2.2 Sustainable development – a global goal

In modern international circles the new concept of 'sustainable development' crops up all over the post-Earth Summit world. Governments have identified it as an overarching environmental goal for society as a whole. The objective has been taken up by many nations in support of the United Nations Rio Declaration, and Agenda 21 treaties have been signed promising detailed implementation at a national level. Government projections involve business as a whole, and individual businesses, changing to achieve sustainable development.

You could be forgiven, however, for not being very clear about what those two words mean, because since the simplicity of the Brundtland Commission's original view, that sustainable development is 'development that meets the needs of the present without compromising the ability of future generations to meet their own needs', there have been many alternative interpretations of what this means in practice.

Implicit in the Brundtland definition are two basic notions. One is a concept of global social justice, implied in 'meeting the needs of the present', which are manifestly still far from being met for four-fifths of humankind. The other is the concept of the conservation or replacement of environmental resources, having specific regard to the needs of future generations.

The environmental economist David Pearce, in *Blueprint for a Green Economy*,[1] gives a summary of the wide range of alternative definitions that are on offer. The meaning of sustainable development is, it seems, coloured by differing conceptions of development as well as moral views. There are different interpretations of the rights of future generations, what responsibilities we have towards today's millions of poor and what formal obligation there is to pass on as good a world to those who follow as we have enjoyed ourselves.

Among these varying views, some see sustainable development as requiring little more than superficial alterations to how we live and do business now; for others, adopting this long-term goal represents an end to current patterns of trading and commercial life as we know it. Indeed, high-profile business leaders are on record as saying that sustainable development is an attractive opportunity for growth. Stephan Schmidheiny, speaking at the launch of the Business Council for Sustainable Development (BCSD), stated that it is 'nothing more than a simple combination of growth and environmental protection'.

Despite this belief in the compatibility between growth and sustainability, the vast majority of individual companies have yet to plan for sustainability. They are far from taking on board the systemic way of thinking required for sustainability to become a practical possibility. In this book we explore what can be done to begin moving towards sustainable operation. Learning to live sustainably is the greatest challenge facing our generation. The reason is quite simple – Planet Earth cannot continue to provide for the rapid expansion of our species and our consumption of such a high percentage of her productive capacity.

2.3 The context: planetary capacity

Business, like all human activity, is a subset of the planet's overall systems. Human civilization is wholly dependent upon the physical infrastructure of the planet, just as the market economy is wholly dependent upon a complex social infrastructure. For most of us in the modern world our relationship with the Earth is exploitative, taking resources regardless of their ability to be regenerated.

The planet's physical systems, on which we depend for survival, have been exhaustively explored and our knowledge base is growing all the time. Within the environmental debate it is possible to differentiate two schools of thought as to how the planet's systems operate as a whole.

Favoured by some environmentalists is the 'Gaia Theory' proposed by James Lovelock, in which the Earth's biosphere works like a self-balancing responsive organism, maintaining a living environment with stable temperature, atmospheric cover and water cycles, creating its own self-adjusting

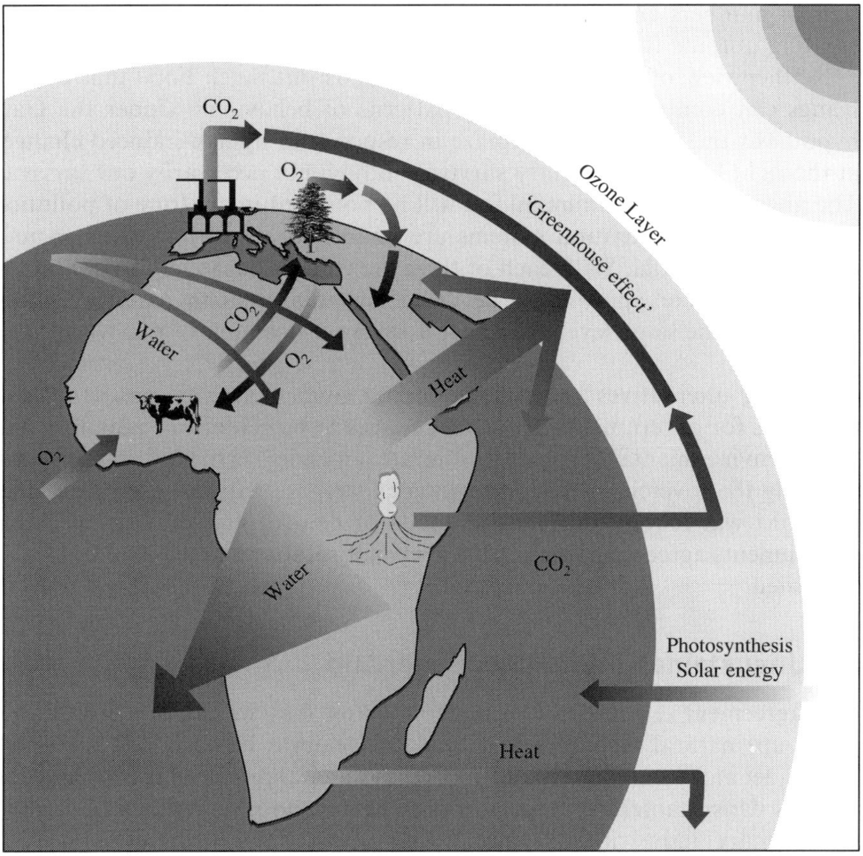

Figure 2.1 Planetary cycles

life-support systems through natural feedback mechanisms (see Figure 2.1).[2] Significant shifts in the climate and structure of the earth, such as those in the Jurassic Age, have happened in the past, and these self-adjustments dramatically alter the environment for life.

The alternative analysis depicts the earth as a 'steady state system' with its own internal equilibrium. It is this equilibrium that is threatened with disintegration through the negative impact of human activities. The first law of thermodynamics states that matter or energy can neither be created nor destroyed. Matter is endlessly recycled, with everything coming from somewhere and going somewhere; matter migrates to other forms, it doesn't just disappear. From this viewpoint the environment's capacity to absorb non-recyclable material is finite. The 'sinks' or absorption mechanisms cannot be indefinitely expanded to accommodate the exponential growth of human numbers and the potential damage our technologies can inflict by creating outputs that cannot be broken down by natural systems.

Neither view of planetary physiology holds out much hope that human beings can continue their existing patterns of behaviour. Under the Gaia hypothesis the earth will restabilize in response to human-induced changes at the right level for planetary survival, though not necessarily our survival. The alternative is that humankind will be poisoned in a latrine of pollution by-products when natural systems are unable to absorb any further non-degradable material. With each of these viewpoints debate continues on the detail, but there appears to be consensus that human activity cannot continue at the same level of impact if severe dislocation of our world is to be avoided.

Finding alternatives to ever expanding human activity poses difficult questions for governments. How are changes to be made? For pollution and other environmental problems migrate outside territorial boundaries, negating the sovereign decision-making capacity of nation states. Providing solutions will require different methods of control and prevention. Most governments agree there is a problem, even if solutions have yet to be clearly identified.

2.4 Unsustainability: a linear scenario

The agreement extends to the understanding that we are still using (and using up) natural resources in a way that cannot be sustained for much longer, let alone indefinitely. Business plays a vital role in that consumption process. Unsustainable practices of present-day global commerce, as well as the primary industries of agriculture, forestry and fisheries which are increasingly becoming profit-driven businesses like any other, include:

- saturation of ‘sinks’, or environmental capacity to absorb pollution and waste
- over-exploitation of most renewable resources
- indiscriminate consumption of non-renewable ones, with little investment in developing renewable alternatives
- irreversible damage to living systems, species, communities, such as those in rainforests or wetlands
- subverting natural equilibrium processes through deforestation, soil erosion, and toxic releases.

Observers who have tried to take an objective view of the effects of industrial expansion, point out that not only are we exceeding the carrying capacity of the planet and consuming non-replaceable capital assets built up over many millions of years, but at present are doing both things faster and faster, as if in a desperate attempt to stretch the natural limits closing in on us.[3]

Taken together, the cumulative effect of continuing these practices into the future is an environment that will not support an acceptable quality of human life, nor a home for other species.

In addition to the degradation of the physical environment, there is a significant threat posed by the current global social system, in which 80 per cent of the world's population stands by and watches the world market economy exploiting *their* environment and *their* resources for the benefit of the wealthier 20 per cent. The gap between rich and poor grows wider, and conditions in many poor countries are becoming more intolerable. Yet politicians, industrialists and ordinary people in the rich one-fifth continue to have faith in models of 'development' which imply that the majority will ultimately reach the standards now enjoyed by the minority.

These models of development have failed on the ground. Individuals who are no longer prepared to wait for development to come to them are becoming increasingly inclined to go off to get some small piece of it for themselves. Given the realities for many developing nations who have suffered actual decline in their standards of living during the 1980s, it is no small wonder that economic migration is increasingly a threat to the fabric of the richer nations.

The future of such extreme economic and social divisions in the global community is hard to envisage. The conventional model of development suggests ever expanding demands for environmental capital. If the goals of development are achieved through this model, try to imagine the environmental impact of multiplying any, *let alone all*, of our present unsustainable practices by five – the apparent implication of bringing the remaining 80 per cent up to our level of resource-intensive development.

When you have completed that exercise, try doubling those anticipated impacts again to cater for future population growth! Or, if that is too grim, imagine the alternative, of accepting, indeed, enforcing, the indefinite continuance of present global inequalities, of 20 per cent of the world population remaining below subsistence level and another 60 per cent being prevented from ever getting anywhere near the living standards of our developed world.

'What if' asks the International Chamber of Commerce, 'within another 35 years there are twice as many people seeking three times as much food and fibre, demanding four times as much energy and engaging in five times as much economic activity – as is projected if the present trends of meeting our expectations continue unchecked?',[4] and their estimates appear quite conservative!

Given these alternative scenarios, the absolute necessity of moving to a form of economic development that can hope to satisfy the needs of the 80 per cent sustainably by reducing the amount of environmental capital required to generate the same degree of economic development becomes inescapable. As we are nowhere near sustainability in the way we provide for even the top 20 per cent at present, achieving a transformation in the way we view both economic and business development becomes a first step of great urgency.[5]

2.5 Drivers of change

As little as five years ago, such statements coming from environmentalists were either greeted with incredulity or ignored. Not any longer! They are now to be heard from the lips of heads of governments and the chief executives of multinational corporations. The people making such statements are among those who have the power to change the very context and ground rules of doing business, and are beginning to use it. There are three forms of power beginning to be harnessed to making business sustainable – *political* power, *economic* power and *competitive* power.

Political power originates with voters and the pressure groups influencing them. Democratic governments follow the political concerns of their electorates, leading to the creation of a growing body of laws, regulations and taxes, bringing changes which inexorably ratchet up the cost of doing business in the old unsustainable way.

Economic power lies in the hands of customers, large and small; from green consumers buying eco-labelled washing machines to public bodies and billion pound companies screening their suppliers' environmental record. The moves by the Clinton administration to screen US Federal procurement using environmental criteria is a clear pointer to this major trend in economic leverage. It is creating new pressures on companies, forcing them to review and find ways to reduce their environmental impacts, 'from cradle to grave' and beyond.

Competitive power, perhaps the most effective of the three, is wielded by successful companies which have taken upon themselves the role of 'environmental leadership' in their sector. They are creating standards and benchmarks of good environmental practice and publicizing them, in order to add to the marketing edge of their products, or subtract from that of their competitors.

So far these things are happening partially, patchily, across the world. But the BCSD, which numbers among its members some of the world's largest corporations, has signalled a growing determination among business leaders to rescue the free market, as well as the planet, by ensuring that in future each business pays and passes on the full environmental cost of its products and processes.

> The cornerstone of sustainable development is a system of open, competitive markets in which prices are made to reflect the costs of environmental as well as other resources.[6]

That sounds simple, but implies an enormous agenda for change. The indications are, however, that it is an agenda which the political and industrial power brokers have now accepted, at least in principle, recognizing that there really is no alternative, other than the collapse of the market system. Radical change towards a more sustainable future is on the agenda from the boardroom to the highest levels of government.

There is a parallel between the progress being made with government and within individual businesses. Environmental issues have changed the understanding of the limits of the nation state and promoted the idea that sovereignty should be surrendered to allow change to happen for the common good as well as at a local level. Likewise, businesses have discovered a need to work collaboratively to secure changes in their environmental performance.

Working together in trade associations, with suppliers and investors, businesses are pooling resources to face these challenges. Just as governments must abide by international agreements, business also finds itself facing common constraints through regulation. Most interestingly, many of the most efficient organizations are encouraging these constraints as a way of ensuring that their dirty neighbours are not given unfair advantage in trading, a similar drive to that of the cleaner nations working to gear up international regulation to their own level of performance.

2.6 Changing the regulatory climate for business

A key debate concerning how business should be regulated centres on internalizing the costs of environmental impacts, thereby forcing integration of environmental protection into organizations' decision making. Traditionally business and society saw nature as a provider of 'free goods' such as water, air and waste disposal facilities. As the limits on the physical environment's ability to provide these services have become clearer, a renegotiation is taking place. Access to environmental goods is being controlled and restricted in response to scarcity.

To date, the means by which users of some of these external resources might pay for their use and repair is undecided, and a number of alternatives have been proposed, using a variety of techniques. From regulation to fiscal measures and the use of market mechanisms such as 'tradable permits', the debate continues as to the optimum pricing mechanism for the enjoyment of facilities such as air and water.[7] Whatever the outcome in terms of particular techniques there is a fundamental shift in progress.

The new relationship between the business organization and the physical world has parallels in the renewal of interest in what constitutes business ethics. Reviewing how external physical resources are utilized questions the nature of individual enterprises and organizations and what responsibilities they have beyond their own boundaries. Companies' boundaries have traditionally been defined in terms of their capital structure and ownership. Current questioning of this view is discussed more fully in Chapter 3. The new tendency is to see companies increasingly as living social structures responsible for their interaction with global society, with purposes other than simply the generation of a return on financial capital.[8] If these new-type companies have an overwhelming objective, it will probably be to become sustainable in the long term.

2.7 The principles of sustainability

The concept of sustainability – organizing our society and economy in a way which does not exceed the long-term 'carrying capacity' of the planet – is simple enough. But how do we translate it into practice? Where do we draw the line? How much consumption, pollution or waste is too much? Do the answers to these questions have to remain a matter for speculation, controversy, guesswork and risk avoidance? Or is it possible to establish and demonstrate simple principles to distinguish between sustainable and unsustainable approaches?

It appears that it is. In an exhaustive two-year process a prominent researcher achieved a nationwide, multi-disciplinary consensus among leading scientists of one highly developed country (Sweden) about what needs to be done. The principle underlying the consensus document is basically very simple – it is to return to a cyclical process of production and consumption.

To achieve this we must relearn the earliest lessons of history. Over a time span of three and a half billion years evolution has transformed a primeval toxic stew into a clean, diverse and ordered biosphere. This has been achieved by living cells processing matter in natural cycles. In the last 200 years, however, we have largely switched from a cyclical to a linear process, by extracting million-year-old mineral deposits and the living components of balanced ecosystems and turning them into growing rubbish mountains, with ten times as much invisible 'molecular rubbish', on the way to saturating our finite atmosphere. What this amounts to is evolution in reverse, turning the ordered biosphere back into the toxic stew!

So what has often been seen as the most mundane and trivial aspect of environmental action – recycling – turns out to be the most fundamental of all, and the simple criterion of sustainability in the long term is the extent to which our economic activities can revert to a cyclical pattern and avoid the creation of by-products which cannot be reprocessed by living cells.

The Natural Step

A remarkable example of national consensus building, The Natural Step, began when Karl-Henrik Robert, a leading cancer researcher, circulated a document on the environmental problems facing society to Sweden's scientific community. That initial draft was revised 21 times until a clear position emerged on the full nature of environmental processes and why things have gone so badly wrong for human society.

This consensus agreed that the cell is the basis of life and how it operates is one of the fundamental lessons of ecology. Cells process matter in self-sustaining cycles, with 'waste' from one species providing nutrition for another in a circular process. For roughly 200 years human beings have been disrupting the cyclical processes of the planet at an accelerating rate, creating wastes that cannot be transformed by natural processes. Some of this is visible and a great deal of it is 'molecular garbage' – consisting of tiny particles which are now at large in the earth's air, water and soil. The volumes are too great for natural systems to assimilate and some molecules consist of compounds that cannot be processed by cells at all.

Armed with this remarkable unanimity between scientists, Dr Robert created an education pack, which was eventually circulated to every household and school in Sweden. Meanwhile fundraising from major companies, organizations and trade unions ensured the financing of the project. The King of Sweden, who is committed to environmental issues, became the patron. In April 1989 the project was officially launched with a nationwide TV broadcast and an environmental gala involving people from the arts and broadcasting. The Natural Step has gone on to build professional networks in many disciplines and is actively working to integrate knowledge of cyclical processes into Sweden's national life.[9]

The American writer Paul Hawken has tried to specify a feasible set of basic requirements for a sustainable business economy. They combine the cyclical principle underlined by The Natural Step, and the principle of cost internalization espoused by environmental economists and the BCSD, with the third prerequisite of sustainable energy. His three 'proposals' seem to us to provide an excellent frame of reference for the future of sustainable business:

1. Build the natural principle that waste equals food into the design of our systems of production, to create a new 'industrial ecology' in which the waste output of one productive process becomes the raw material input to another.

2. Create a system of 'feedback and accountability', by means of green fees and taxes and new public utilities, to ensure that ecological costs become part of financial costs, so that business can grow rich by restoring the environment rather than destroying it.
3. As part of the movement from living on capital to living on income, and to address the urgent need to slow down climate change, Hawken proposes moving from an economy based on carbon (in the form of fossil fuels) to one based on hydrogen and sunshine – both potentially inexhaustible energy sources (at least in any human time scale), if we are prepared to invest enough of our existing 'capital' in developing them for practical application.[10]

2.8 Sustainability and the individual business

The emerging consensus on the need for sustainability, and even on its meaning, does not of itself tell individual managers what to do about their business now. The available range of strategies, as we show in Chapter 4, remains wide – from ignoring the signs and going for short-term advantage, to explicitly making sustainability the strategic focus for the future of the business.

But what sustainability as a concept does for the individual manager is to place a large number of disjointed pressures – eco-management and audit, pollution control, waste management, recycling, energy efficiency, green consumers, CFCs, purchasing policy, life-cycle analysis and more – into the coherent framework of a major historical development. The move towards sustainability is comparable in scope with the first industrial revolution, involving shifts in behaviour which will inevitably transform the way we do business from the ground up.

> In ten years the market will be about nothing but sustainability. Whether you are a business man or a Florence Nightingale, the answer is the same: close the cycles and you will be the winner tomorrow. (Karl-Henrik Robert.)

The value of sustainability as an overall aim is to formulate a clear and rational objective for activity. All subsequent action, and the application of processes and techniques, are therefore stages in a progression, allowing clear analysis and a sense of achievement. We feel that there are some very persuasive reasons for companies in key sectors to embark on a strategy of sustainable development sooner rather than later, and these are discussed in Chapter 4 in comparison with the other strategic choices available.

It must be admitted, however, that for the present such a decision entails much more than choosing to do X rather than Y. While a decision to conduct an environmental audit or to go for BS 7750 certification is a fairly straightforward matter of following well-documented paths, a strategy of sustainable development has to be invented almost from scratch. There are no benchmarks to emulate; this will be a process of discovery.

The international business community has responded with the International Chamber of Commerce (ICC)'s Business Charter for Sustainable Development as an introduction to the overall agenda. Others would challenge the underlying assumptions behind these principles, and an alternative vision has been articulated as the CERES, formerly Valdez, Principles. The full versions of these documents have been included in Appendix 1 and you may like to compare the alternative attitudes to 'environmental capital' and the issue of sustainability that they display.

Helpful though such calls to action are, they do not and cannot lay out the strategic goals appropriate for an individual organization nor the means required to make such goals reality. What is required is a more specific focus.

2.9 Criteria for sustainable business strategy

In any strategy development process there will be a wide range of issues to be addressed which relate to particular circumstances and sectoral issues. Nevertheless it is possible to spell out some of the general criteria for deciding whether a strategy of sustainable development is genuine.

The company positively planning for sustainable development is:

1. Attempting to measure the extent of its present *un*sustainability in terms of 'true' environmental cost, calculating what are the 'remediation' costs, for repair or replacement, of the environmental capital being used (see Chapter 5 on Financial Strategy for an expansion of this question).
2. Developing an alternative set of environmental accounts with the aim of 'internalizing' its environmental costs, initially on paper and later in reality.
3. Identifying for each unsustainable element of the business:
   - best estimates of environmental/remediation costs
   - feasibility of alternatives
   - how critical the element is for the business overall
   - competitor reactions and industry trends.
4. Planning specific actions, including investment, to move towards sustainability in different aspects of the business.
5. Exploring the possibilities of developing cyclical processes and 'industrial ecosystems' in collaboration with other organizations.
6. Coordinating the competitive and financial strategies that drive the business from day to day with its plans for reaching sustainability.
7. Linking the environmental, economic, social and human aspects of

corporate performance as parts of a system able to operate sustainably over decades.

8. Working towards living off the income from human-created capital resources rather than ecological capital, software rather than hardware, where profit comes from ingenuity and adding value, from creativity rather than using up physical resources without funding regeneration.
9. Attempting to regain its own economic security by becoming as economically independent as possible, generating financial capital from within its own resources or working with investors who view capital accumulation as the priority rather than short-term gains or immediate dividends.
10. Developing constructive dialogue between regulators and the business, as such relationships will change out of necessity as sustainability targets are set for the overall development of the economy.
11. Working to educate and inform its stakeholders of the issues concerning sustainability.

In general we believe that a company's commitment to sustainability will be manifested through:

- the products or services supplied
- the processes through which these are generated and how they are managed, monitored and controlled
- organizational structure, ownership and financial strategy of the company or business
- the quality of relationships with stakeholders.

2.10 The relationship between business output and society's needs

When considering sustainability as a fundamental business objective, as opposed to environmental improvement, we believe that *what* is produced by a business is as significant as *how* it is produced. The relationship between the products and services supplied by an enterprise and the capacity to move towards sustainability is a complex one.

Environmentalists and business were mutually antagonistic for many years and criticism of the environmental impacts of business led to defensive reactions from the business community. Historically, when challenged the answer to charges of pollution and degradation was 'Society wants the products we produce and therefore it must be prepared to pay for the environmental damage'. Moving beyond this negative image of industry as an unscrupulous environmental exploiter requires a more positive vision, something beyond simple confidence in the social value of short-term 'wealth creation'. For some critics this popular phrase just appears to mean the annexation of natural wealth by the rich from the poor and by those now living from those not yet born.

Ultimately today's consumer preference and choice in the developed countries cannot be the sole determinant of the validity of the role of business in wider society. To achieve a fuller understanding it is necessary to ask whether business should be *meeting wants* or *answering needs*. The difference is not simply one of semantics; no debate about the further direction of industrial society can shy away from the issue of demand *creation*, when reviewing questions regarding the use and replacement of scarce natural resources.

The fundamental human needs have been defined as the material conditions that will provide food, shelter, work and a social setting which safeguards our physical and emotional health, supporting our ability to procreate and to be social beings.[11] In consumer societies there is constant encouragement to experience the fulfilment of wants for which we have to be enticed to hunger, through the creative use of advertising. The advertising industry creates complex messages, which associate purchasing with our desires for status, power, autonomy or more basic emotions.

From an environmental perspective the use of precious physical resources to satisfy a seemingly endless stream of artificially created wants is difficult to justify. There are many products that meet true needs; the debate is whether encouraging people to constantly change their belongings, often before the end of their productive life, is appropriate. In many cases manufacturers also have a role to play in designing for repair and continued usefulness rather than sales being dependent on everything being scrapped before its time.

The context for any escalation of material demands is that the costs of ecological capital, as well as financial and manufactured capital, are bound to increase over the coming decades due to scarcity with increased pressure on environmental carrying capacity. Activity in the world economy is also likely to increase as more nations and people try to enter the open liberalized market. These relentless pressures must throw into relief the way in which resources are consumed and to what ends.

The debate on demand generation and the environment focuses on the polarities between a perceived 'freedom' to have whatever is potentially available and perceptions of 'responsibility' to consume less to allow a more equitable distribution of resources within society and across the world. Currently there is no credible political force to focus attention on how commerce might be motivated to satisfy human needs above wants, but this dilemma will need to be resolved at some point in the pursuit of sustainable development. In the meantime for the individual company, however, reviewing the relationship between its product offering and overall human welfare forms part of a strategic approach to sustainability.

2.11 The context for corporate environmental strategy

In this chapter we have identified changing expectations within global society which in the medium term, some 10 to 20 years, will create basal shifts in attitudes. Exactly how such diverse pressures will manifest themselves will vary according to country, the speed of environmental decline and willingness of the international community to address that challenge over the coming decade.

For the individual business, as a matter of course, reacting to these changes in the long term will have to be accommodated alongside immediate business issues of survival, competition and development. Therefore, having a clear understanding of how the global operating environment is being changed by environmental pressure is essential. Without this strategic perspective there is a real danger of allowing immediate operational pressures to postpone the search for sustainability until it is too late.

References and further reading

1. D.W. Pearce, A. Markardya and E.B. Barbier, *Blueprint for a Green Economy*.
2. James Lovelock, *Gaia – A New Look at Life on Earth*, Oxford University Press, 1987.
3. Paul Hawken, *The Ecology of Commerce*, HarperCollins, 1993, p.23.
4. Jan-Olaf Willums and Ulrich Golüke, *From Ideas to Action: Business and Sustainable Development*, ICC, 1992.
5. See Alexander King and Bertrand Schneider, *The First Global Revolution: A Report by the Council for the Club of Rome*, Simon & Schuster, London, 1991, for a full account of the global 'problematique'.
6. Stephan Schmidheiny with the Business Council for Sustainable Development, *Changing Course*, MIT, 1992, Chapter 2. The chapter is preceded by a quotation from von Weizsäcker (see note 6):

    Bureaucratic socialism collapsed because it did not allow prices to tell the economic truth. Market economy may ruin the environment and ultimately itself if prices are not allowed to tell the ecological truth.
7. See Ernst von Weizsäcker and Jochen Jesinghaus, *Ecological Tax Reform*, Zed Books, 1992. We would recommend this small volume for a short résumé of the best options for business friendly and revenue-neutral changes in tax regimes.
8. Charles Handy, *The Empty Raincoat*, Hutchinson, 1993.
9. *Current Sweden* No. 401, The Swedish Institute, December 1993.
10. P. Hawken op. cit.
11. The most famous expression of this is Maslow's Hierarchy of Needs; for a more comprehensive statement see the works of Manfred Max-Neef as featured in Paul Ekins, *Wealth Beyond Measure: An Atlas of New Economics*, Gaia Books, 1992.

3

Ecological management – beyond the boundaries

1. This century companies have been forced to take on a widening range of social responsibilities.
2. At the same time the method of 'bounded rationality' – limiting the factors considered in decision making – has dominated management.
3. Sustainability requires a more systemic approach incorporating all the potential 'stakeholders'. The boundaries of the system to be managed have to be widened.
4. Citizen concern about a variety of health, environmental and ethical issues has produced new pressure groups.
5. These represent the latest extension to the stakeholders of the business.
6. The emerging model of management includes a vastly expanded set of stakeholders most of whom have an interest in the organization's environmental strategy.
7. The prevailing approach is still one of piling additional constraints onto the bounded rationality model, manipulating costs and paybacks from the outside. It assumes that managers will behave in a primitively selfish rather than a professional and responsible way. To achieve sustainability this model needs to be replaced by the systemic approach of ecological management.

The post-war period has seen an unprecedented growth in the scale and turnover of industry and commerce in the industrialized world, an omnipresence generating widespread awareness of the potential impacts, for good and ill. With growing sophistication business assumed the mantle of social responsibility by claiming the role of 'wealth creation', while critics are increasingly concerned over the supranational powers of global companies and their role in increasing inequality.

Within this context the environment is becoming something of a test case. Can business be trusted when it claims to be working voluntarily and proactively towards the solution of one of the biggest problems the world has faced? Or will it continue to need coercing (by legislation) or bribing (by tax incentives) to take its environmental responsibilities seriously? At the level of the individual business this ambivalence manifests itself in a clash of values, with the environment seen as a constraint, an opportunity or, less frequently, a moral imperative. Whether activity is proactive, driven by an internal dynamic, or reactive, dependent on external stimuli, is critical to environmental performance.

3.1 Changing obligations

The structure and purpose of the corporate organization is constantly evolving, as the relationship between individual enterprises and society at large changes to incorporate an ever developing range of responsibilities and expectations.

In retrospect it seems hard to imagine the lack of constraints on businesses earlier in this century. In the industrialized north things have certainly changed from the days when inhuman working conditions were commonplace. In part, this reflects a change in society's overall value system and power balance, as well as a shift in the outlook of business owners and managers.

The pace of change was forced by the pressure of organized labour, through trade union militancy and by the skilful use of labour shortages. External pressures arose from mobilized consumer groups, local communities and national single-issue campaigns, bringing the negative social and health impacts of industrial activity to the attention of a wider audience. Demographics and the fluctuating relationship of women with the world of employment also played its part, as women again entered the paid workforce necessitating a renegotiation of organizations' obligations. The interweaving of the state and individual enterprises in many Western countries during the Second World War and subsequent creation of welfare systems also left a legacy of companies administering substantial social provision.

These changes are formalized in the legal infrastructure in which business operates. The web of obligations has been expanded over decades. Many companies are now required through taxation systems to administer social welfare provisions in the form of sick pay and pensions, to provide healthy and safe working conditions and to ensure that consumers are protected.

Day-to-day operations are covered by tax laws, contract laws, labour and employment laws, monopoly laws, health and safety laws, equal opportunities laws, information disclosure and, finally, environmental laws. These obligations fall on the company in its capacity as a separate legal entity and the obligation to ensure implementation of these commitments is the personal responsibility of its directors and managers.

Concern for health, financial integrity, safety, quality and consistency are part of the framework of best practice in a multitude of ways, and detailed regulation ensures both that these obligations are complied with and that public scrutiny is possible. The relationship of an individual business, and its directors and managers, with those outside the company is therefore much more complex than the often quoted single duty to create the largest possible financial return for the owners of the enterprise.

Of necessity, the company expresses its relationships to the outside world through the actions and voices of the men and women who are *de facto* the company at the time of operation. It is how these individuals view the world and their responsibilities towards it that holds the key to changing the impacts of human activity on the environmental systems that sustain life.

3.2 Boundaries – a management tool

The principle which underlies most modern management techniques, and which is probably the main influence on the way in which managers think about the world and take decisions, is the principle of *bounded rationality*.

Bounded rationality consists of putting a notional boundary around the entity being managed and focusing attention within it on a unitary goal (such as maximizing earnings per share or minimizing unit cost) subject to a set of *constraints* (such as legislation, resources, costs). Intellectually this is a very powerful approach which gets results. It is comparable in some ways with the method of experimental science – which has also been phenomenally powerful and successful – of changing one variable at a time while controlling the rest.

This process is based on the highly practical and understandable idea of optimizing an individual part by ignoring the whole – because the whole is too impossibly complex to cope with. This is practical up to a point; it breaks down, however, when the factors pushed outside the boundary in the effort to achieve a limited objective superbly catch up with us and we find that by succeeding extraordinarily well at one level we have failed extraordinarily badly at the next, more encompassing, level.

There is a growing suspicion that this may be the point we have reached in our economic history, 'the failure of success' as it has been called. The fear is that the growing level of disfunction we create as a by-product now outweighs the benefits of the original process, though it is hardly popular to point this out.

The mental process of bounded rationality is undoubtedly a powerful technique, yet it is just that, a technique for achieving an objective – efficient management of an economic process. When looking at management in the last years of the twentieth century, does this technique deliver the optimum results? We believe it has begun to fail, and nowhere is this more apparent than in the response to the environment.

An alternative approach has been evolving this century, one which we have to take seriously if we want to be successful in managing the impacts of enterprises on the natural environment. The alternative consists of progressively widening our boundaries and looking at more than one variable or one objective at a time. Managing by monitoring a range of variables represents a more *systemic* and, in the last resort, *ecological* approach to management.

3.3 Beyond the limits of bounded rationality

The systemic approach to thinking about an organization's responsibility has been developing in parallel to the growth of its legal obligations. In the history of management theory this century, the notional boundaries of the systems to be managed have been pushed further and further beyond shareholders' equity, to include first the employees (Human Relations theory), then the human impacts of technology (Sociotechnical Systems), the local and national community (Corporate Responsibility), customers (Total Quality Management) and suppliers (Supply Chain Management).

The incorporation of systems theory in both management and earth sciences has opened the way to the development of theories both around global ecology and ecological management, going beyond the linear methods and limited objectives of bounded rationality.

Using an 'Open Systems' model borrowed from biological science, some writers have redefined the task of management as regulating a set of cyclic flows – of inputs and outputs, benefits and obligations – between the organization and all its 'stakeholders', in practice adding customers, suppliers, vendors, distributors and bankers, the local community and the nation-state to the shareholders and the employees.

The conception of the entity to be managed is expanding to incorporate this whole system of flows, with management taking on the role of optimizing, improving or enhancing a whole set of performance criteria in a balanced way.[1] Figure 3.1 outlines some of the key relationships in this system.

Just as we learnt that controlling individual emissions does not guarantee improvements in the overall environment, as one pollutant could be reduced only to be replaced by another, thus in business management introducing the stakeholder concept involves consideration of a *portfolio* of input–output exchanges to be balanced for optimum success.

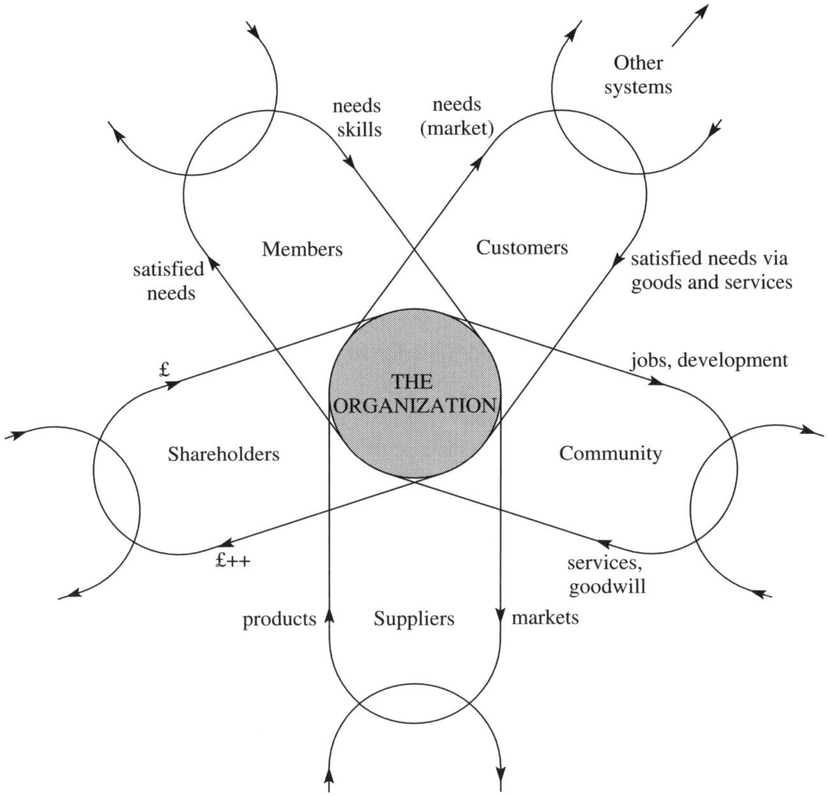

Figure 3.1 Stakeholder system
Source: Alan Dale, *What is Organizational Development*?, Brunel Management Programme.

3.4 The stakeholder concept

The stakeholders in an organization are the groups, beings or systems that are affected by the company's operations and whose support is required to safeguard the company's future.

Awareness of stakeholders beyond the obvious participants in an organization grew in the 1980s, as the extent of industrial society's impacts on the global environment became more widely known. Health and environmental concerns showed some of the most dramatic examples of external pressure on corporate practice. Companies operating within the following markets were forced to take account of new stakeholders in their enterprises, as consumers made it clear that they were unwilling to accept certain levels of risk:

- Concerns about the threats to human and global health from the depletion of the ozone layer led to massive consumer reaction against CFC-based aerosols over a very short period of time.
- Similar concern about the potential for human health damage from cattle infected with bovine spongiform encephalopathy (BSE) led to controls on agricultural feedstock producers in the UK and a drop in beef consumption.
- Fears about the origin of timber for DIY stores led to supplier auditing procedures as firms struggled to identify whether their materials were from sustainable sources.
- Worries about the levels of heavy metals in fish stocks in various parts of the world led to new information about areas being fished and the nature of commercial fishing. Increased knowledge raised the issue of the incidental killing of dolphins by tuna fishermen, which led supermarkets across the world to monitor the harvesting of this food source with more care.
- The battery-producing industry responded to consumer concern about heavy metal pollution with projects to recycle spent batteries and investment in removing the most offensive of the metals involved, such as cadmium.

There are many examples of small-scale pressure groups having raised uncomfortable issues for organizations through the diligent pursuit of information, and nowhere have they been more successful than in the environmental arena.

The addition of collective groups to the list of stakeholders, in the form of local community and consumer organizations, brought with it a change in the relationship between enterprises and their public. The creation of recognizable groups able to articulate the concerns and demands of ordinary citizens for their own well-being and that of other communities and environments, gave organizations real faces across the table to which to respond, requiring rational arguments based on demonstrable progress, as well as 'class actions' to fight in the courts. (We discuss in depth the impact of stakeholders on corporate communications in Chapter 11.)

The coalescence of groups is an ongoing process and one which continuously modifies the range of stakeholder concerns that managers have to respond to. Advocacy groups develop constantly, with the aim of representing the views and needs of diverse communities and environments affected by commercial and industrial activity. So-called 'active' citizenship, the advent of high-powered and affordable personal computing, high-

Table 3.1

Company	*Stakeholders*	*Stakeholder needs*
	Shareholders	Maintenance of welfare or risk minimization
	Parent company	Continuation of benefits (e.g. income, supplies, job opportunities, wealth creation, community involvement)
	Senior managers	Compliance with laws and other sanctions
	Directors	Control and remediation of environmental damage
	Employees	Continued integrity of natural systems
	Trade unions	Information
	Customers	Continued capacity/existence
	Competitors	
	Suppliers	
	Vendors/distributors/franchisees	
	Financiers	
	Insurers	
	Local community and economy	
	Local and international environments	
	The international community	
	Regulators	
	Government	

quality reproduction capacity from consumer electronics and global communication networks are all moving the organization towards a twenty-first-century scenario where new, vocal stakeholders appear on the scene even for such silent victims as distant communities, marginalized sections of society and non-human species who cannot speak for themselves.

Once manifest, stakeholders have a variety of needs – for information, for continued benefits, for an 'absence of disbenefits'. The role of corporate management now is to reconcile the conflicting needs and requirements of this wider group of stakeholders, balancing the needs of each to ensure optimum satisfaction (see Table 3.1).

The relationships of this set of stakeholders with the organization provides the extended boundaries of what currently constitutes the individual business. In terms of this model, 'managing' the business includes the joint optimization of all these relationships and achieving a balance of satisfaction between the stakeholders, subject to the legislative and ethical climate of the day. Managing interaction with, and optimizing the satisfaction of, stakeholders is going to be the key to business management in the next century.

One of the greatest changes this new approach to management involves is seeing that the future is itself a stakeholder in the business. This requires a shift in thinking and a significant reappraisal of objectives, but living this change is a precondition of achieving sustainability.

When did environmental stakeholders become an issue for business? Local concerns about polluting practices are centuries old. It was the extension of knowledge during the 1970s and 1980s of the dynamics of environmental impacts beyond the locality and 'smoke-stack' to wider impacts upon the global climate, oceans and air quality that brought about a qualitative change in the relationship between individual firms and industrial society. Stakeholders in businesses around the world began to assert themselves, in many cases demanding information that had previously been hidden or not deemed important enough to collect. Information disclosure and transparency of working practices is increasingly becoming society's expectation when it comes to environmental concerns. These raised levels of expectation represent a major shift in the terms of legitimacy for business. Without feedback on environmental performance, automatic acceptance of either competence or corporate morality can no longer be taken for granted.

3.5 Identifying the stakeholders

Stakeholders are an increasingly important dynamic of business. The questions for any corporate strategist are:

- Who are the stakeholders?
- Have their needs been audited?
- What strategy has been adopted to meet their concerns or to manage a constructive dialogue with them?
- How can their satisfaction be maximized overall?

The following represents a broad cross-section of potential stakeholders to be considered.

Shareholders

Having stressed that shareholders are not the only stakeholders, we should not forget that the providers of capital to the business, both large and small, share a need to maintain the financial integrity of the enterprise to ensure continued financial benefits in terms of both capital value of their investment and dividends. Understanding the environmental strategy of the company, and the implications this will have for long-term performance, is crucial to meeting both needs. Beyond performance, communication and long-term analysis skills are required to optimize shareholder satisfaction as a full stakeholder. In addition, managers may increasingly have to take account of shareholders' interest in their company's ethical stance, as the ethical investment movement gathers momentum.

Parent company

Like the shareholders, the holding companies of many groups demand continued benefits in terms of financial success and the minimum risk possible. Meeting both requires awareness of the strategic dimension of environmental issues and new measures of performance that can incorporate achievement in this field into group-compatible results. (Further discussion of these issues follows in Chapter 5 on Financial Strategy.) Strategic vision, measurement criteria and feedback on performance will be needed for optimization.

The potential environmental liabilities inherent in mergers, acquisitions and sales of assets will need to be considered. Consistency is also important as each territory has its own legal constraints on clean-up and remediation. Corporate policies based on working to the highest standards across all locations will, if implemented, require that such obligations are not undertaken lightly.

In a group the overall portfolio of investment holdings should also be assessed for its environmental risk potential, as one under-performing operation can have significant public relations and financial consequences for other operations. Each operation needs to have access to adequate operating guidelines.

Senior managers and the board of directors

As the individuals who are *de facto* the company in operation, their commitment to overall environmental goals is of critical importance. Most modern environmental legislation contains a provision for criminal liability of directors and other senior officers of companies, who may be personally convicted where their company commits an offence which is attributable to their consent or neglect. This personal liability also covers health and safety and fire regulations. Risk minimization is just one of the strategic responses that are available.

Real stewardship and strategic management requires analysis of the sustainability of the organization and the development of a transitional strategy and its implementation. Given the increasing tendency to reward directors and managers with share options, enlightened self-interest should also ensure that the long-term health and commercial viability of the organization prompts attention to these matters (though in the short term the effect may be different).

Employees

The individuals who make the business function on a day-to-day basis probably have the biggest stake in it. Employees have a complex set of needs – for continued employment, a healthy company in the long term (particularly for those with pension entitlements), high safety standards and, since some staff members may well also be shareholders, a secure financial return on their investment.

The workforce will also be implementing and managing the environmental improvements in performance that are required to make the company sustainable, and to achieve these goals motivation and commitment will be required. This demands a well-executed awareness building programme and good change management skills (see Chapter 10 on environmental awareness training and change management).

It should also be noted that the trade union movement is beginning to understand the significance and importance of environmental issues for the well-being of their members. Active trade union pressure is being applied in some countries and sectors to ensure that standards are improved and the long-term interest of all members of the company is considered.

Customers

In consumer and business-to-business markets the customer has begun to actively influence corporate strategy by favouring companies which have environmentally-friendlier products and can demonstrate a responsible attitude to the environment. Supplier auditing is increasingly being used to determine the environmental impact of products used by major industrial purchasers and government departments and agencies (see Chapter 9). Customers, especially for products which involve long-term commitments to a particular technology, service back-up, spare parts or even brand values, require the company to be viable in the long term.

Competitors

Competitive advantage is a clear and coherent reason for investment in environmental performance, and gaining an insight into the strategy of one's competitors is vitally important. For some organizations leadership of their sector is part of their mission, and to achieve a change in working practices within their industry they are prepared to invest in educational and cooperative ventures to improve overall standards.

The Chemical Industry's support of the Responsible Care Programme originally developed in Canada,[2] is one of the clearest examples of the extent to which companies have a stake in their competitors' environmental performance. The absence of competitors' environmental action is experienced as an inhibitor, their activity as a spur, to a company's own environmental strategy, but in addition, when a whole industry's reputation is affected by the performance of the laggards, the industry finds itself threatened with unwelcome attention from the regulators.

IBM and ICL's support of post-use machine recycling, as well as Thomas Cook's and British Airways' endowment of the World Travel and Tourism Research Centre at Oxford Brookes University,[3] are other examples of working with competitors. In North America the support given by more than 20 leading companies to the Global Environmental Management Initiative (GEMI) projects, especially those regarding benchmarking, such as the Environmental Self-Assessment Program, demonstrate a willingness to 'promote a world-wide business ethic for environmental management and sustainable development, to improve the environmental performance of business through example and leadership'.[4]

Suppliers

Suppliers are major stakeholders in the company and the relationship is a symbiotic one. For organizations which live by trading or retailing products manufactured and assembled elsewhere, suppliers are critical to the pursuit of environmental improvement. Understanding the environmental impacts of product lines is simply not possible in such circumstances without the support of suppliers. Exploring what changes are feasible, and then ensuring that changes are actually made, depends on the mobilization of these third parties.

Large companies in monopolistic positions may be able to unilaterally impose environmental buying criteria, but it is a high-risk strategy. Coercion should never play a part in relationships with such important supporting organizations, where a partnership approach is important. Facilitating suppliers' awareness, acceptance of and commitment to the challenge of sustainability is a critically important factor in the fulfilment of a strategic shift towards sustainable business.

Vendors/distributors/franchisees

For manufacturers, their relationship with retailers and distributors is of vital importance in allowing the environmental dimension of the product to reach consumers. If the relevance and complexity of the environmental benefits are not understood by retailers and distributors, consumers may never receive the information on which to make a reasoned judgement.

Long-term, managing product life cycles and improved reliability would suggest changes in the relationship with consumers, as manufacturers build closer ties with the end-user and undertake end-use disposal systems such as take-back schemes and technology upgrading.[5] With this shift in emphasis between manufacturers and sales outlets in prospect, maintaining and developing improved relationships with manufacturers will also be in the interest of retailers.

Financiers/insurers

The ability to maintain good relations with the financial support networks on which a business depends is of critical importance. There was a time when bankers did not have to concern themselves with the environmental and social performance of the companies to which they lent money and insurers happily took their annual premiums secure in the knowledge that everything was under control. Environmental disasters, contaminated land and contingent liability have put paid to this professional distance. Increasingly more institutions recognize that risk management extends to understanding the environmental dimensions of their portfolios and now demand more information on operational performance.

Local environment

In times past companies polluted without thought of the consequences for the local environment, and as we learn more about recent activities behind the old 'Iron Curtain' we are reminded of the consequences for the local environment in dirt and loss of visual amenity, levels of contaminants and the health risks involved. Local pollution in the West is less visible, but can have equally devastating consequences, through the build-up of radioactivity in an area or dioxin contamination, methane concentration and chemical spillages.

Beyond controlling negatives such as pollution, some companies try to positively enhance the area in which they work by providing support for local schools, community groups and using their sites to increase habitats for indigenous wildlife or providing facilities for the less fortunate. Improvements in amenity on site can also have beneficial results for the whole local community. A further step is for companies to collaborate with other organizations, both public and private, in a coordinated plan to protect, improve and restore the environment of a locality or region, working through a Regional Environmental Management System (REMS).[6]

Local communities and economies

In the same way that the local environment suffers from a careless approach to its welfare, local communities and economies can also deteriorate under a regime of neglect. A single organization may be extremely important to the well-being of its particular location, providing a high proportion of work and underpinning large numbers of subcontractors and service companies. The productive capacity of the local community which feeds the organization can be strengthened by its involvement in the welfare of the community, either directly through charitable donations or indirectly through facilitating staff involvement.

Adopting a policy of local supply wherever possible not only makes sense in transport costs but also in environmental and community terms. Allowing the local community to retain as high a percentage as possible of the income generated provides for a more vibrant and living community for staff, managers and visitors to enjoy. The alternative is a cycle of decline and forced relocation, as the company's original sites deteriorate in value through the decline of the surrounding area. Viewing the local community as a strategic asset makes corporate growth and development both more likely and less costly.

Regulators

Regulators have a major stake in the organization's ability to comply with laws and regulation, and often wish to upgrade its capacity to meet the regulatory challenges rather than just engage in clean-up operations. Finding out what is expected of the company is the minimum requirement of any environmental scoping study, but talking to the regulators about the best available technologies to solve particular problems occurs less often. Many regulatory bodies are sympathetic to companies with a genuine commitment to improvement and have data on less generally known technologies and leading edge techniques.

Government

Government is a stakeholder in the nation's businesses as it receives benefits in the form of revenues and the distribution of social welfare provision. Governments across the world have committed their populations to targets on sustainability and reducing environmental damage. Increasingly business is seen as both an agent for this change and a source of technological solutions to policy problems.

The international community

The international community is also a stakeholder, and here we mean the peoples of the world, not just their governments. As relationships between organizations trading across the world have become so complex, it is often difficult to remember that there is another human being at the end of the deal. Many people in developing countries are living in conditions that the average Western family pet would refuse to exchange, and ensuring that commodity and other producers in the developing world receive adequate returns for their labour and production is one of the challenges of the next century. It will not be possible to improve environmental standards globally if large parts of the world cannot earn enough to feed themselves, living decently and sustainably. Fair trade is an issue that sits uneasily with the passion for liberalization seen in the Uruguay Round of the GATT process, but it is an issue of enormous importance to genuine sustainable development.

In response, some companies are developing bilateral trading relationships with particular groups of producers to ensure that more of the value of the commodities purchased stays within the originating community. Buying brazil nuts to use in hair conditioner or ice cream may not be able to overturn the global trading system, but such gestures by leading companies, as well as producer-distributed products such as Café Direct (a fair trade coffee brand in the UK), do point a way forward to changing the human and ecological impacts of trade. As companies dealing in Third World manufactured carpets have found out, deciding who makes one's stock can be an important strategic decision, if boycotts and consumer campaigning on child labour are to be avoided.

The international environment

Like the carpet traders and retailers operating blissfully unaware of the reality behind their products, before the hole over the Antarctic was discovered many manufacturers and users of CFCs (though not all – see Chapter 8) were unaware of the potential impact on the ozone layer of these previously considered stable gases. Obviously knowledge moves on and some of us will be caught unawares, but since the understanding of global-scale damage to the planet's ecology has become widely known there is less excuse for complacency. The major environmental issues for the next 20 years were introduced in Chapter 2, and a strategic (i.e. long-term, thought-through) approach will be required by every business, as well as every institution and every community, if we are to make the necessary changes needed to meet these challenges.

In an ecological stakeholder model managers should realize that *the natural environment itself* has become one of their major stakeholders, as much as, if not more than, any shareholder, customer or bank. Like them, the environment provides the business with contributions it cannot do without, like them, it can benefit or be damaged by the way the business is managed, like them, it has the power to close the business down. It is not necessary to personify the environment, as is done in the 'Gaia Hypothesis' (see Chapter 2) to understand the truth of these relationships.

3.6 Ecological management: an end to boundaries?

Everyday experience shows that for the present 'bounded rationality' is alive and well at the heart of Western culture. The continuing dominance of this approach over political and commercial policy is holding back the evolution of sustainable lifestyles and business systems. It leads to the conviction that managers need to be *forced* to consider environmental issues by the application of outside interventions, either through legislation or financial penalties and rewards. It is assumed that with such limited governmental interference undesirable effects such as pollution and the destruction of natural resources can be brought back from that no-man's land of 'externalities' (the economists' term for what is left outside the boundary) to be included in the decision-making process as costs or constraints, to make ending global damage an automatic outcome of traditional methods of decision making.

This perceived need to 'price' anti-environmental behaviours has led to the new discipline of 'Environmental Economics', in which a range of solutions are proposed through taxation and incentives which would allow managers to continue in their mission of 'wealth creation' but in a way which incorporates the environment in the calculation of economic costs and benefits. In the UK Professor David Pearce has been the most high-profile of the economists proposing techniques to be utilized to adapt the existing systems to new circumstances.

Similarly, the new specialism of 'Environmental Law' is rapidly documenting new regulations and control mechanisms to restrict noxious emissions, to regulate the movement and use of harmful substances in industrial processes, to arbitrate in cases where damage has occurred and redress is sought and to control international disputes over resources.

These two approaches follow the prevailing linear pattern of policy making, with the state in theory operating outside the stakeholder debate, arbitrating and ensuring sensible outcomes. The apparent implication is that industrialists are either incapable or unwilling to adapt autonomously to the new facts of a threatened natural environment, but have to be manipulated by a set of sanctions; that changes in the behaviour of organizations can only be achieved by politicians, lawyers and economists instituting a new set of rules for managers to play by.

Those who are active in day-to-day business dealings will recognize only too well the fallacy in presuming leadership of business from the outside. Industry is frequently in the lead, with politicians trailing the opinion formers of commerce, and never more so than in the current climate of deregulation and joint partnership between state and private capital. Moreover, we do not find ourselves in a stable situation where all industrial processes and products *and* their environmental effects are well enough known to be controllable by law. Technological development is speeding up, not slowing down, and while the industrialists who are introducing new processes and materials may not know enough about their environmental effects, they still know a lot more than legislators who often only lock stable doors after horses have bolted.

Yet, in spite of these uncertainties, wider society needs to be able to hold business to account. Both sides are facing a real dilemma. Business has learnt how to operate successfully within the present bounded framework, and many managers are unwilling to go further than accommodating to whatever laws and taxes they have been unable to head off by determined lobbying. The majority of people in industrial societies want the market economy to continue because of the goods it delivers, but many do not trust the motives of those who run it and want all industry and commerce to be made to behave responsibly. To date, the only mechanisms under consideration for achieving this control have been the law, taxation and 'the market'.

There are many who clearly understand that the current situation is inefficient and sub-optimal. Almost certainly, motivated and intelligent managers could do a better job for their company and for society if they were enabled to take full responsibility for their environmental impacts. Almost certainly, society would have a better industry and a better environment if it was able to approach this problem as requiring a real responsibility to find solutions, turning the search for a sustainable economy into a collaborative enterprise based on mutual trust. Such a process poses the questions of the time scale involved and how and when sanctions should be used to ensure that the unscrupulous and the laggards are not allowed to profit from their selfish exploitation of the natural environment.

Until that external debate is resolved realistically, the only management answer we can see on the horizon is through the emergence of what we would call *ecological management*, taking the progression charted in this chapter to its logical end – a model of management which extends the boundaries of managers' responsibility to take in *all* the potential effects of their decisions, and which seeks to optimize not a single restricted performance criterion, such as profit, on behalf of a single class of stakeholder, but the balance of benefits for *all* the stakeholders who contribute to and are affected by the organization's outputs and performance in *all* their aspects.[7] A professionalism based on these principles just might create the basis of trust for business and society to collaborate more creatively than at present in building a sustainable world economy, while law making, new performance measures and suitable financial inducements can be used as a back-up to ensure commitment and motivation in the transitional period.

References and further reading

1. A fuller account can be found in K. Knight, *Ecological Management: Towards a New Model*, an unpublished paper obtainable from the author.
2. Stephan Schmidheiny, *Changing Course*, MIT, 1992, pages 221-4.
3. *World Travel & Tourism Environment Review* 1993, World Travel & Tourism Council.
4. *Environmental Self Assessment Program*, Global Environmental Management Initiative, Washington DC, 1992.
5. Tim Cooper, 'The Durability of Consumer Durables', *Business Strategy and the Environment*, Vol 3, Part 1, Spring 1994.
6. R. Welford and A. Gouldson, *Environmental Management and Business Strategy*, Pitman, 1993.
7. See Note 1. The term 'ecological management' has also been used by Richard Welford, to describe an approach to management 'which has its roots in deep ecology and involves a shift to a holistic worldview.' R. Welford, *Environmental Strategy and Sustainable Development*, Routledge, 1995.

PART TWO
Foundations of environmental strategy

In Part Two we lay out the main strategic dimension of a corporate approach to the environment. At present the environment is experienced by most companies as only one pressure among many, and not necessarily the most urgent one at that. To gauge the true long-term significance of this apparently new phenomenon for the future of the company demands stepping back from immediate concerns and viewing the business in a broad perspective of global change.

In so doing, one of the most persistent obstacles to any radical reassessment is the apparent constraint of what is or is not economically viable. Often, however, current financial orthodoxy and more or less arbitrary classification of income and expenditure streams, inhibit us from doing full justice to changing realities. This is why it is so vital that a truly strategic approach to the financial dimension should be brought in at the start of any review.

The two chapters in this Part address the corporate strategy agenda.

Chapter 4 sets out the broad strategic choices open to the company and looks at how they can be actualized through the corporate strategic planning process.

Chapter 5 reviews the ways in which financial management can underpin a business environmental strategy directed towards sustainable operation.

4 Strategic options and the planning process

1. There is an urgent need to include the environment in companies' strategic planning.
2. In the long term sustainability will be a condition of survival for every company. In the short term, however, this is not self-evident to all, and a number of different strategic responses have emerged, including:
   the do-nothing strategy
   the defensive strategy
   social responsibility
   strategic opportunity
   planning for sustainable development.
3. So far the majority of firms have not integrated their environmental and business strategies.
4. A company's environmental approach can make an important contribution to its business development strategy.
5. This involves making assumptions about the future development of the business context. There are a number of possible scenarios, described under the following headings:
   Business as usual
   High turbulence
   Technological fix
   National and global interventionism
   'Darwinian' evolution
   Voluntary sustainable development.
6. Development of strategy options for an organization will benefit from a process that enables a wide range of inputs from different functions to be worked through collectively, both analysing existing knowledge and exploring future possibilities. We present a range of approaches for doing this, including:

Context assessment
Environmental challenge focus
Consensus building
Benchmarking
Delphi
Advisory panel of experts
Tracking social attitudes
Scenario planning.

Recent research suggests that companies are likely to move through a number of stages of strategic development on the way to fully incorporating the environment in their corporate strategy.

We must no longer wait for tomorrow, it has to be invented. (Gaston Berger, French philosopher.)

4.1 The need for planning

To recapitulate: the ecological equilibrium of the earth is in imminent danger, in not one but manifold ways, endangering the general health and well-being of people, animals and plants. This threat to the natural environment is having a powerful impact on the business environment. Companies find themselves being held accountable for their conduct by a growing set of stakeholders, which include, ultimately, the natural environment itself.

The demand, widely supported, that they should in future internalize their environmental costs by paying for the pollution and resource depletion they cause, is putting in doubt the profitability, and even viability, of many industries. The question of how to achieve sustainable development concerns not only economists and governments but every business that aims to survive into the twenty-first century.

In consequence, companies now have to include the environment in their strategic planning for the future of the business. Issues of the magnitude of those raised by the environmental agenda necessitate a strategic view; their inclusion in strategy development and the corporate planning process is vital. Monitoring and responding to the anticipated changes in society, as it moves towards a more sustainable way of living, will require a wide spectrum of planning inputs and projections to increase the likelihood that any resulting environmental investments are both commercially viable and represent responses of long-term significance.

Planning for business development in general requires a knowledge of the context in which current operations take place, decisions on the most influential trends requiring a response and a model of strategic development, to harness this information and guide operational changes in response. In the move towards environmentally sustainable operations each business has to develop and integrate two visions: one of its own individual future and one of the overall changes likely to take place in the society of which it is part.

It is not evident that many existing companies have taken their thinking about environmental strategy as far as this. For most, environmental management policy is as yet a rather secondary issue, a response to extraneous pressures far from their core concerns. Certain 'underlying' environmental strategies can be deduced from companies' overt statements and observable behaviour. But at present these are more likely to be expressions of corporate character, value system and stage of strategic development than the outcome of careful strategic analysis.

The argument of this chapter, and indeed of the whole book, is that the environmental issues now facing business are of sufficient seriousness and complexity, and sufficiently vital to the survival of individual enterprises, to require a more detailed analysis.

In the latter part of this chapter we look at the ways in which environmental pressures may affect accepted business development strategies, and review some of the alternative macro-economic scenarios for a society adjusting to the environmental agenda. We also focus on the strategic planning process itself and look at specific environmentally relevant techniques to assist in developing and choosing between strategic options. Finally, we review a model of how organizations develop strategically as their knowledge of the environmental agenda evolves.

Before embarking on this exploration of strategic environmental planning, however, it is useful to outline the major 'underlying' or generic approaches to environmental strategy which companies are displaying, whether by conscious choice or as a reflex reaction to events, and which represent the gamut of strategic options available.

4.2 Strategic options for management: how does the individual business meet the challenge?

Within the wider challenge to our society of building a sustainable market economy, each business faces its own private sustainability problem. For, if the world economy as a whole is to survive within a tolerably preserved natural environment, it may only do so through a classical Darwinian process, in which companies that are able to adapt to the changed requirements of a 'sustainable world' will displace those unregenerate polluters and resource guzzlers who are unable to meet the environmental cost of their depredations. The lack, at present, of any credible political alternative to the market economy makes this the only serious conclusion for business and political leaders, and raises the question for the individual company whether it has the makings of a dodo or a survivor.

In the short term, however, the outlook is not nearly as clear as this. Of the regulatory, fiscal, financial and commercial processes, and the changes in management outlook, which will bring about the transition to sustainability, only a few are as yet in place. Many of the old mechanisms for externalizing environmental costs are still operating, and it may appear to some industrialists that rather than it being the case that 'the polluter pays',[1] the situation is still one in which 'it pays to pollute'.

The 'free lunch' of bargain basement energy, and material prices based on the cost of extraction, have enabled many businesses to make profits from what are in essence uneconomic operations, when viewed from a sustainability perspective. In the short term there may appear, for individual businesses, to be no alternative to continuing in the same pattern. And even when, as a society, we are faced with incontrovertible evidence that our present way of operating cannot possibly continue, many of us do not really believe it, deep down; or we put our faith in the technological fix that will alleviate the need for painful changes in behaviour.

Perhaps this is why it is possible at present to distinguish a fairly wide range of strategic responses to the environmental question that is being put to business – from doing little or nothing and resisting government intervention, to seizing on the challenges of sustainability as a heaven-sent business opportunity – and why not every business is either aware or consistent in its response.

The 'do nothing' strategy

There are at least three versions of an environmentally minimalist stance.

A) 'FLAVOUR OF THE MONTH'

Some managers still see the environment as a passing fad that will go away if they keep their heads down, a transitory bandwagon for consultants, careerists and publicity seekers, a cash cow for academics and quangos. (There is even a small group of writers, proponents of the 'greenhouse conspiracy theory', who claim that the world is really moving towards a new Ice Age, and that the global warming hypothesis is a scientific plot to extract unlimited research funds from governments.)

The fact that there is probably a small grain of truth in many of these jaundiced views should not blind anyone to the massive amounts of wishful self-deception underlying them. Yet many business people still think this way, even if they no longer feel comfortable saying so against the growing environmental consensus. As a strategic position 'Flavour of the month' has the advantage of helping to keep your eye on the ball in the short term, but that is about all. In the evolutionary stakes it seems fated to extinction.

B) NOTHING TO DO WITH US

A very widespread version of 'Do nothing' is the belief that environmental protection matters – for *other* companies. As for us, runs the argument, our firm is too small/does not use enough chemicals/ is not the kind of business to affect the environment.

As regards size, small and medium-sized companies (SMEs) account for a substantial proportion of European Union business and a significant share of pollution. Moreover, with the growing adoption of green supplier policies by many large companies, SMEs will find it increasingly difficult to maintain a 'Do nothing' stance if they wish to keep their customers.

A company does not have to use toxic chemicals to damage the environment. Businesses in the financial sector, for example, often confine their environmental concern to recycled paper and lead-free petrol, while their lending and investment policies can be the drivers for massive environmental destruction elsewhere. As the BCSD points out: 'Scarce capital can be attracted to enterprises or projects that undermine rather than promote sustainable development.'[2] Despite the significance of these impacts, many financial institutions still view their lending and investment policies as remote from the corporate environmental agenda.[3]

Most non-manufacturing business activities have environmental impacts, from distribution and transport undertakings which contribute to pollution and climate change by their fuel usage, to offices with their energy use, space heating, air conditioning and equipment choice. As the environmentalist Jonathon Porritt put it at a seminar: 'To be is to pollute'.

C) FIGHTING FOR SURVIVAL

Many companies say that fighting for the survival of their business through recession or market turbulence leaves no time or attention for any but the most short-term concerns. Environmental protection, like staff training or product R and D, is seen as a luxury that cannot be afforded for the time being.

'Luxuries' can, however, very quickly turn into necessities with even small changes of the business and regulatory climate which alter the cost–benefit ratio of environmental policies. And arguments based on 'survival' can only too easily conceal the fact that profits have actually become dependent on the company being able to export its emissions to the community's air, land and water free of charge. The relative cost structure of any industry based on cheap energy or waste disposal will be rapidly transformed by the introduction of taxes to correct those externalities, such as carbon taxes or landfill levies. Any business based on the free availability of natural resources has no long-term viability, and the short-term 'survival' strategy no more than puts off the evil day.

While such short-termism may be an 'economically rational' strategy, to maximize profit from an unsustainable business while it lasts, in order to fund sustainable development elsewhere, the intangible costs in reputation and goodwill must not be overlooked. A recent survey of *Fortune* magazine's top 500 US corporations shows that two-thirds of those listed in 1956 have disappeared. The common characteristic of the survivors is a strong moral code: these are companies who take care of their record, rather than simply trying to present a massaged image.[4]

Defensive posture

An alternative form of short-termism acknowledges the power of the environmental pressures from government, community and the business world itself, while remaining profoundly sceptical or opposed towards them. It states that the business of business is to 'create wealth', and anything that gets in the way of this single-minded pursuit is to be avoided. As a strategy this means finding ways of neutralizing environmental pressures and preventing them from interfering with the main thrust of the business.

A) REACTIVE COMPLIANCE

The simplest approach is to limit one's environmental policy to one of compliance with the law. Like health and safety, contracts of employment and the Factory Acts, observance of the relevant regulations is made the responsibility of appropriate specialists (often those catch-all Health and Safety Officers, with Environment tacked onto their title), who are expected to mediate between operational and commercial needs and the requirements of the regulators.

Such a strategy of compliance without commitment is often primarily one of damage limitation, it is commercially and ethically tenable in the short term. Adopting such a strategy however does not exclude fines for exceeding consents and appearances in the green groups' lists of polluters, perhaps because the HSE Officers charged with ensuring compliance are not always the most powerful of figures and they do not always get their way. (In certain industrial sectors, however, such as chemicals and nuclear energy, health and safety are given very high priority and can strengthen environmental enforcement at the operational level.)

But the very idea that there is a need for 'damage limitation' (meaning damage caused to the firm rather than by it) comes from a view of the environmental issue as a threat to business. If this was a correct perception, its implications would be profoundly depressing. It would mean, in essence, that the environmentalist critique of capitalism is actually valid: that there is no way in which the present economic system can coexist, long term, with a viable planet.

B) RESISTANCE

The more active version of the defensive strategy is to resist all new environmental proposals for as long as possible. The intensive lobbying which foiled the European Commission's first attempt to get agreement on the principle of carbon taxes before the Rio Conference is only one recent example of this. There has been a consistent history of orchestrated business resistance to all the main proposals for environmental improvement during the 1980s and 1990s.

The scientific case for any environmental protection proposal is never 100 per cent water-tight, in what is, after all, an area of new research. So the first line of defence is to produce some contradictory findings and call for more research. The next barricade is claiming the impossibility of modifying industrial processes within anything like an acceptable time scale, and pointing to the threat of mass bankruptcy and unemployment if it is attempted. The third line of defence is the spectre of becoming uncompetitive if any changes are made before all other countries have agreed to do likewise. Finally, there is the threat of having productive capacity moved from the present territory, as the Dutch government found to its cost.

When, finally, the popular and political pressures become too powerful to resist any longer, valuable time will have been lost and irretrievable damage may have been done. A recent case study from Germany showed how concerted industry resistance over a five-year period to changes in power generation led to a window of opportunity in the refitting of a number of major power plants being missed, to the cost of both the objecting firms and the economy (see Example 4.1 below).

C) RISK MANAGEMENT

At the more sophisticated end of the spectrum, what started as a short-term defensive stance is turning, under the pressure of legislation, customers and public opinion, into a full-blown *risk management strategy*. Such a strategy is difficult to distinguish, in its outward manifestations, from the more

Missed opportunity

Ulrich Steger describes an example of the economic damage that can be done by industry resistance to environmental measures. The first draft of the German large combustion plant regulation, concerned particularly with eliminating SO_x and NO_x emissions from power stations, was tabled in 1977, but, owing largely to the resistance put up by industry, did not become law until 1983. By that time, the intervening build-up of popular pressure resulted in a much tighter deadline than might have been possible earlier, and the industry was given until mid-1988 to comply. This gave very little leeway for technical development, and the industry had to use available (end-of-pipe) clean-up technology and pass over more promising 'clean technology' solutions which were not sufficiently developed. The retrofit of clean-up equipment coincided with the modernization of a large proportion of the generating equipment, effectively 'fixing' the sub-optimal solution until the year 2000, and leaving very little scope for the more efficient technology for eliminating the pollutants from the process. Steger comments: 'Economically this not only meant a missed innovation opportunity, but also a reduced capital productivity with higher running costs.'[5]

Example 4.1 Missed opportunity

proactive forms catalogued below, since risk management strategy is based on the dawning realization, now increasingly widespread, that environmental failures can be as disastrous as commercial or technical blunders, and as hard to recover from. A full-blown risk management approach to the environment (as described e.g. by Newman and Breeden[6]) guards not only against technical and scientific risks from major incidents and shortfalls, but also considers the relations of the company to its significant stakeholders, all of which can be damaged by environmental factors, with associated effects on the performance of the business. It means keeping an eagle eye on changes in consumer buying habits and customers' supply-chain policies, as much as on regulatory changes being mooted in the European Union (EU) and new environmental criteria being considered by insurers, lenders and institutional shareholders. The risk management strategy seeks to monitor stakeholder concern and contingent liabilities and to use them as a yardstick for the investment payback of environmental protection measures.

Social responsibility

A) CORPORATE CITIZENSHIP

A contrasting strategy towards the environment is found among companies with a long tradition of emphasizing corporate social responsibility. Many such companies will have environmental policies of long standing, which were initially an expression of their corporate citizenship, side by side with other employee and community care policies. Such companies have long defined their business strategy at least partly in terms of solidarity with, rather than exploitation of, their social and natural surroundings.

In the present climate of high environmental concern, such companies place an enhanced emphasis on their own 'green' policies and initiatives and look for ways of turning them to corporate account. In some cases they see it as their mission to raise standards and give a lead to other companies, while at other times they develop new views of strategic opportunity which can help them to capitalize on a high environmental profile.

B) RAISING STANDARDS

As part of a strategy of corporate environmental commitment, such companies are embarking on policies of continuous improvement, involving targets for emission control, waste reduction, recycling and use of recycled materials, energy efficiency, technological and product improvement, environmental screening of suppliers and customer education.

In raising their own standards of performance, at the same time they set new benchmarks for their competitors and for public regulators – new standards and market entry barriers from which they will be the first to benefit. Such programmes seem to be found particularly among market leaders – they are supported by good commercial performance in the first instance and are used in turn to support further enhanced performance.

C) ENVIRONMENTAL LEADERSHIP

In the same social responsibility vein, some companies commit themselves to a policy of environmental leadership. This means a willingness to 'stand up and be counted' among those who take environmental protection seriously, going public about their green initiatives and not being afraid to 'show their head above the parapet', even at the risk of being shot at by green activists for something they have done or omitted to do. Some go a step further by actively campaigning for environmental and related causes, like The Body Shop, which uses its high street sites and transport vehicles to display well-designed campaigning messages.

Strategic opportunity

The idea that, far from being a threat to their business, the environmental crisis can actually represent a once-in-a-lifetime opportunity to build competitive advantage, increase market share and even open up new markets, underlies the strategies of some of the most visibly pro-environment companies.

A) COMPETITIVE EDGE

The possibility of gaining competitive edge by offering 'environment-friendlier' products dawned on a number of manufacturers of consumer goods during the first wave of green consumerism in 1989/90. One of the early successes, widely publicized, was Fort Sterling's 'Nouvelle' brand of recycled toilet tissue (see Chapter 7 for further examples). But while some manufacturers took the trouble to make genuine changes to products, others focused more on changes in their advertising. This led to some predictable cynicism among consumers and environmental groups, exemplified by Friends of the Earth instituting their 'Green Con Awards'.

Not surprisingly, with growing recession during the early 1990s, marketing strategies that tried to gain competitive edge as well as a price premium, coupled with consumer confusion about the genuineness of green claims, the green consumer wave peaked and subsided, apparently confirming the flavour of the month views of hard-bitten marketers. Nevertheless consumer surveys continued to show unabated interest in greener products, and work continued on an EU-wide ecolabelling scheme.

In the meantime a new arena for green competitive edge loomed on the horizon. As well as consumers, business customers started to take an interest in the environmental credentials of their suppliers. One of the earliest shock waves in the UK market came when B & Q, the DIY arm of the Kingfisher group, started sending out a long and detailed environmental questionnaire to all their suppliers. (See Chapter 9 for a fuller discussion of these developments.)

B) IMAGE BUILDING

Although corporate image advertising has a long history and many sceptics state 'It's like wetting your pants – no one knows you're doing it but it gives you a nice warm feeling', companies with a public commitment to responsible corporate citizenship have long provided examples of the competitive advantage to be gained from a favourable corporate image, over and above the public perception of individual brands. Whatever degree of scepticism there may be concerning such corporate spending, many millions are spent every year in producing corporate communications and increasingly these include environmental issues.

A new version of this strategy is to use one's environmental performance as the basis of a major profile-building PR campaign, as has been done by Norsk Hydro UK, a company involved in such sensitive areas as chemical fertilizers and salmon-farming – a good example of grasping the strategic opportunity inherent in the environmental agenda. The decision to pioneer environmental reporting significantly raised the profile of the UK subsidiary of this European corporation, allowing it to join the ranks of corporate giants whenever the environmental issues regarding business are discussed.

C) CLEAN-UP AND CLEAN TECHNOLOGY

In purely commercial terms, by far the largest area of business opportunity is, of course, clean technology. In 1990 the UK Centre for the Exploitation of Science and Technology (CEST) estimated the UK market for environmental protection at US$140 billion, the European market at up to US$860 billion and the US market at up to US$1060 billion for the decade of the 1990s. (See Chapter 7 for further information.) There are really two markets here – in the first instance for cleaning up existing polluting processes by what are commonly described as 'end-of-pipe' devices, and beyond this a market for radically new technologies which avoid pollution at source and utilize materials more effectively.

There is a potential opportunity here not only for plant manufacturers but for any manufacturing company investing money to clean up their own processes by development of technology which can also be sold to others (see Chapter 7). Environmental 'leaders' have a good chance to score here, while the 'defenders' and 'resisters' may have to pay the price of coming last. For example, the development of new clean processes in the electronics industry has led to the creation of satellite companies to market the new technologies and skills.

D) INNOVATION

It has also been suggested that the environment, like any major challenge, can be used by companies as a strategic opportunity to build up or renew their innovative capability as such. The packaging industry is seen by many as a prime target for the wrath of the environmentalist. There are, however, companies in that sector with the strategic vision to see environmental issues as a significant opportunity. The demand for light, recyclable, solvent-free packaging materials, able to withstand modern handling processes yet aiding easy disposal, allows for the creation of niche, and ultimately mainstream, markets making available new materials to satisfy the demands of environmentally responsible manufacturers and retailers.

Sustainable business

Although a number of companies mention sustainability in their environmental policies, few have set out to achieve it in any systematic or comprehensive way.[7] Even *Changing Course*, the report of the BCSD, while it develops lines of policy and accumulates international examples of good practice all aimed at environmental improvement, does not attempt to say how much improvement is needed, nor what targets business ought to aim for.

Similarly, attempts to provide standards and guidelines, the British Standard for Environmental Management Systems (BS 7750), the Guidelines for Environmental Performance Measurement[8] and for Environmental Purchasing Policies[9] produced by Business in the Environment (UK), only provide standards and guidelines for the *systems and procedures* which companies are to use in establishing and implementing their policies, without attempting to identify what needs to be achieved.

The general directions towards sustainability are known, but which are the specific paths to take and how far should we follow them? Any attempt at a full-blown sustainable business strategy has to try to answer these questions for the company concerned. In so doing, it will come up against the central dilemma of sustainability – the principle is clear, but no one can tell with certainty where it begins or ends in practice.

Sustainability is not a fixed grail to strive for, but something we have to invent. And 'we' means all of us, politicians, managers, farmers, voters and consumers, as much as scientists, economists or environmentalists. Nowadays none of us in the industrial world (and very few outside it) have any experience of what it means to live or work or make money sustainably.

Which poses the questions whether in these circumstances it would not be foolhardy for any company to attempt a strategy of genuine sustainable development (as distinct from a sort of green opportunism – publicizing its green achievements and keeping quiet about its shortfalls)? Should it not wait for some authoritative guidance from the politicians or the scientists? We think not.

Because business's environmental actions have arisen largely as a reaction to outside pressure so far, many managers believe that any future actions should similarly be based on the demands of the law-makers, the scientists, the consumers. This is to overlook the crucial transition that is taking place before our eyes, right now. It is business people who are saying to governments in response to BATNEEC[10] type initiatives: 'Don't prescribe the processes and technologies we are to use, but tell us the outcomes you want us to achieve and leave us to find the most effective methods.' It is business people who are declaring:

> 'Given the large technological and productive capacity of business, any progress towards sustainable development requires its active leadership.' (S. Schmidheiny with the BCSD, *Changing Course*, 1992.)

The fact is that there is a lot of work to be done to discover in practice what measures businesses must take in order to become sustainable, and it is work that only businesses can do after having made an active commitment to a policy of sustainable business operation. Admittedly, they will be doing the pioneering work on behalf of the more cautious companies who are hanging back to see which way the dog will jump. But by the same token they will put themselves ahead of the game and in a position to become the business leaders of the 2000s.

The gap between aspirations and actions is large at present. The academics developing the indispensable new systems of environmental accounting which sustainable development requires cannot find companies willing to commit resources to working collaboratively with them, in order to put these systems to the test. Yet the first company in an industrial sector to have developed a working environmental accounting system for itself will have acquired a head start in the quest for combining sustainability with profits.

So what might a sustainable business strategy entail? In Chapter 2 (section 2.11) we set out some of the criteria for deciding whether such a strategy is genuine. They include identifying and systematizing the measurement of current unsustainability, assessing costs and alternatives, planning, investment and coordination with day-to-day strategies.

Hence embarking on a 'sustainable' strategy means doing more than creating a shopping list of feasible and affordable environmental improvements; it requires looking systematically at the environmental impact of all operations and calculating the 'remediation costs', i.e. the cost of putting back the environment as you found it, or the equivalent.

That information, once it has been consolidated, provides a starting point for action, which will consist, over however long it takes (within short-term survival constraints), of scaling down, modifying, substituting or, at the extreme, eliminating, individual operations until an affordable remediation cost has been reached and implemented for each.

Unsustainable use of renewable resources has to be dealt with by finding ways of cutting back until some point of transition between unsustainable and sustainable consumption has been reached, having first estimated where that point lies. The use of non-renewable resources can be tackled head-on by planning for substitution by renewables within the time scale of exhaustion of the resource.

Beyond the need to address sustainable production there are fundamental questions to be asked about the nature of what is produced and whether the business is, in essence, sound. Do the goods or services produced support a sustainable lifestyle for others? Can sustainable delivery or production processes still create the margin required to make the business viable, even when environmental impacts have been fully costed and paid for?

These are only some of the more obvious implications of opting for a strategy of sustainable business. There will be many more. It is obvious that such a strategy is not for the faint-hearted. But then, probably the same can be said for any proactive business strategy, or, indeed, for the decision to stay in business at all, under the foreseeable conditions of the twenty-first century.

4.3 Integration of business and environmental strategy

To maximize the opportunities offered by environmental issues, individual businesses must integrate their environmental knowledge with their business planning. Current research suggests that very few companies have embedded the environment into their strategic processes.

In 1993 the Environmental Management Centre at Brunel University in the UK conducted a survey of major companies' strategic responses to the issue of environmental protection. Most companies responding recognized the importance of the environmental challenge to their operations and did not see a conflict with their overall business strategy. A majority were judged to be more reactive than proactive in their current approach and were not attempting to integrate the environment into business or competitive strategy. The remainder were including a 'green' focus in some of their corporate strategies, but this was not always supported by strategies for innovation and corporate response to competitors. It is also clear that at the present time financial and human resource strategies have yet to include the full implications of implementing overall corporate aims for environmental excellence.[11]

But for those companies who recognize the importance of integrating the environment into their core business strategies, how can this recognition be used to shape the alternative approaches open to them for developing their business itself?

4.4 Business development strategies: the environmental opportunity

Current approaches to business development in companies rarely take account of the environmental agenda. Responding to competitive and external pressures calls for detachment from the company's existing operating practices and focus, and for the ability to look for opportunities to use resources and situations differently. When reviewing alternative options, the classical routes forward are:

- Innovation
- Diversification
- Integration
- Internationalism
- Cooperation.

Each one of these avenues may be significantly changed by an appreciation of the environmental dimension.

Innovation

The development of new products and services in environmentally friendlier forms, reducing the environmental impacts of producers and consumers, will be critical in the move towards making the economy more sustainable. Widespread recognition of this strengthens the business case for an innovation strategy and can help to guide its direction. Selecting the strategic innovation option, however, first requires analysis of both the internal capacity to provide new solutions and the availability of adequate financial resources to implement the development programme.

Innovation potential varies according to the individual company's corporate culture and the dominant commercial culture in which it is placed. There is a variance between the approaches of Anglo-Saxon organizations and German and Japanese companies, with their greater emphasis on quality and engineering product improvements.

The ability and inclination for innovation in response to the environmental pressures may be affected by these factors. Acceptance of the need for sustainability will, however, for many organizations and sectors, require a strategic commitment to innovation if change is to be implemented within the timetables set by both legislation and the impending breakdown of many natural cycles and support systems. In some sectors such as automobiles, refrigeration and water supply, a global race is already underway to find and implement the new technologies that the environmental agenda necessitates. Increasing the speed with which new offerings are developed is vital to the provision of environmental goods and services, both for the long-term health of the originating businesses and also for their customers.

Diversification and integration

The classical choice when seeking expansion, and one much exercised during the 1980s in the Anglo-Saxon economies, is that of diversification or integration. There is a multitude of examples, from that of Firestone and other multinational companies buying rubber plantations in the 1920s to Sony's 1990 purchase of Columbia Pictures Entertainment, seeking to secure software for its consumer electronics interests. These multinationals have chosen to diversify and integrate to protect their sources of raw materials supply or their potential for market growth.

The rise of the conglomerate through the amassing of unrelated business operations, chosen for asset value or income potential, has changed business structures in the post-war period. The creation of such mixed business portfolios required sophistication in management thinking. This led to the creation of the Boston Consulting Group's famous analysis, with its pseudo-financial, money-dealing language. Portfolio analysis offers a rationale for the development of organizations on such a scale. With the pressure of environmental issues liable to impact on company value and cause repercussions across the whole operating group, today's trend towards devolution of power to smaller operating units does not by itself reduce the risk inherent in holding such a wide range of interests.

Evaluating the differing environmental potential of companies across a portfolio highlights the need to manage the move towards sustainability carefully, working slowly and progressively, improving one operation after another. From the wider perspective of a group of companies the costs of a strategic shift of this nature has to be managed within an appropriate time frame and disposals made if necessary, as investment levels are adjusted overall. Assets of some unsustainable operations with a 'cash cow' status can, as we have seen, be utilized constructively to allow cross-funding for the establishment and growth of more sustainable alternatives, benefiting the whole group over time. The quest for corporate environmental excellence may, therefore, include the use of diversification and integration as a significant contributor to development strategies.

The potential macro-environmental scenarios outlined below will entail risks, to both supply lines and markets. For example, the threats posed to the biological stocks on which many businesses are built may promote backward integration to protect supplies. As water levels fluctuate due to global warming impacts reducing water availability, companies owning shares in water irrigation facilities are more likely to survive than those solely dependent on the public systems. Similarly, the pressure on marine stocks may force companies dependent upon declining fish populations to enter the fish-farming industry to maintain supplies.

A current example is that of the fine musical instruments industry in the UK which is finding problems in maintaining its traditional sources of timber. Many of the trees that were feedstocks for that industry are now endangered and are only exported from their originating countries under special arrangements. Companies used to importing timbers freely are now working with charities, governments and local groups to attempt to protect their traditional materials. The alternative is to lose the ability to make classical instruments in fine woods.

Internationalism or decentralization

Expanding horizons by setting up business in another territory is a classic manoeuvre. The need to address environmental issues is both a constraint and a driver for such geographic changes. Sustainable technologies are much sought after; rare skills or product inputs are likely to be pursued wherever they are available.

As with integration moves to protect market power, the need to find alternative uses of capacity when home markets decline may also ensure a more international perspective. Less positively, there is also the prospect of companies being unwilling or unable to make the transition to more sustainable ways of operating, moving operations to territories with less legislative pressure so that they may operate 'dirty' processes profitably while trading freely.

Environmental issues also exert another powerful pull in the direction of reducing size and localizing production, through the increasing cost and environmental impact of energy use. There are inefficiencies inherent in many large-scale processes, and global specialization based on cheap transport may become severely constrained when the environmental costs of long-distance transport have to be internalized by industry. Implementing strategic changes of the order that will be called for may also be easier in smaller units, where the innovation and value shifts required to underpin sustainable business may be more manageable.

Cooperation

Unlike traditional strategies based on adversarial models, there is a business strategy based on collaboration, which may be of particular relevance in the face of the environmental threats to our economic system. Where the challenges ahead are too great for the individual business to resolve, many organizations are opting to work with their competitors, suppliers, industry bodies and even environmental groups to find solutions to matters of joint concern.

A positive role for suppliers in business strategies often reflects earlier commitments to long-term relationships. Preferred supplier status has built up mutual confidence which can make joint problem-solving a worthwhile investment. Competitors with parallel strategies, for example those based on niche rationales or with separate geographic aspirations, have much to gain from working with another company to share the financial and resource implications of change. Environmental management benchmarking is one example of the new-found mutuality developing between major players as they reappraise their strategic opportunities.

4.5 Business planning in what context? – Alternative futures

Strategic planning involves making assumptions about the future. One approach to this is through the technique of scenario planning which we discuss below. To indicate some of the current assumptions – implicit or explicit – which underlie thinking about the developing impact of the environment on business it may be helpful to outline a representative set of approximate 'scenarios'. These are by no means mutually exclusive and some of them may even be complementary.

Business as usual

This is based on the view that environmental change is really nothing special, just one of the many challenges to business operations that are regularly thrown up over time. There is faith in the inherent flexibility of the capitalist system to overcome such problems. Unlike views which anticipate major dislocation, there is no assumption that the overall principles of operation have to change. Instead, it is considered that the steps needed will be incremental and can readily be accommodated. Technological innovations in this scenario are largely confined to end-of-pipe clean-up technologies rather than leading to significant changes in manufacturing processes.

The 'Business as usual' approach is still one component of UK government and EU policy making, where some of the more difficult questions about increasing material consumption, transport needs and pollution from energy generation continue to be soft-pedalled. Environmental degradation is seen as a business opportunity as much as a threat, as end-of-pipe technologies are exportable to communities with industrial inheritances of dirty technologies, such as those in Eastern Europe. If these attitudes prevail, the growth sectors will be clean-up technologies and environmental campaigning, as clean-up depends on regulatory standards being set!

High turbulence

This scenario represents a pessimistic view of the outcome of businesses and governments persisting in a 'Business as usual' approach, leaving environmental discipline broadly to unregulated market forces. In this picture there are pockets of sustainable operation in the midst of social and economic turbulence. The lack of a considered political response to the issues brought up by the collapse of natural systems leaves no-one 'in charge'. Environmental degradation apparent through impoverished local environments, loss of ozone layer protection and changes in the global weather system due to global warming lead to extreme conditions and related social stress.

Support structures for the business community such as the global insurance markets are reduced, as risk management becomes too problematic. There is an increase in the overall financial stakes of doing business as potential liabilities escalate. Business opportunities abound at a time of chaos, as needs are also escalating. Chaotic market signals require short response times and the ability to transform structures rapidly. Political turbulence and uncertainty make investment planning difficult.

Technological fix

In contrast, this scenario is based on the belief that technical innovation will provide all the answers to our environmental problems. Here science and technology, 'experts' of all sorts, corporate research and development, between them hold the key to the future. This view holds out great opportunities for business, and major markets to be won in providing technological solutions to environmental problems. There is little certainty, however, of a smooth transition from the 'post-industrial' to the environmental technology age, given the inevitable problems caused by the irregular development cycle of innovation, with the bottlenecks that will have to be worked through until new solutions are found. There is evidence that this scenario is influencing some Japanese thinking.

National and global interventionism

This scenario envisages environmental protection being enforced by a combination of legislation and economic instruments; environmental risk is therefore contained by the power of governmental intervention harnessed to market mechanisms. Within this framework the international competitive market economy is seen as the most efficient mechanism for change, through harnessing the power of macro-economic signals such as resource pricing. The most eloquent example of this view is found in the BCSD's book, *Changing Course*.

Much discussion has taken place of the need for an internationally level playing field in which clean businesses are not unfairly disadvantaged by the lack of investment and quite literal environmental 'dumping' of their less respectable competitors. Just as the risks are recognized to be global, so, in this scenario, minimum standards of commercial behaviour are expected to be internationalized. Under these conditions the largest organizations prosper, as they are better able to both lobby for and afford the requisite level of regulation than their smaller, less rich competitors. Growth is found in regulatory enforcement, and among large companies and organizations with controlled environmental impacts and, therefore, lower compliance costs. Innovative smaller companies producing new solutions are subject to takeover by predatory, larger, well-funded groups.

'Darwinian' evolution

An extension of the interventionist scenario is the possibility already indicated, i.e. the development of a sustainable economy by a form of evolutionary pressure comparable to that described by Darwin. The survival of the fittest, i.e. the least polluting, means that markets are taken over by sustainable companies, designed for organic growth and able to feed off the share lost by their non-sustainable competitors whose backward practices make their costs too high in an environmentally regulated marketplace. There is extensive carnage among unsustainable firms, while healthy corporate growth is possible for the survivors, once the investment in sustainability has been paid for and long-term viability guaranteed through sustainable structures and practices.

Voluntary sustainable development

In this, the most optimistic and orderly of the scenarios, the need for sustainability is widely recognized within society. There is agreement on the need for action, both commercially and politically, with governments setting targets for change and industry being left to implement organic development. This 'corporatist' view envisages a cascade of policy from government, through to industry bodies, going on to implementation at the level of individual firms.

Within this context individual companies with leadership aspirations lobby governments, aiming to influence the style, timing and approach to be adopted. Corporate action is framed by acceptance of radical strategic change, rather than attempting to block activity. The response to Greenpeace's challenge to the insurance industry to lobby for action on global warming[12] represents an example of this approach in action. Opportunities for business would abound from the institutional demand unleashed by governmental policy truly committed to sustainability objectives, as each government department and local government seeks to implement low-resource use, energy efficiency and closed systems management.

There are few examples of national governments actively developing the vision required to launch such activity within the commercial world, although the Dutch, with their National Environmental Plans, do appear to have moved some way towards the governmental dimension of such a process.[13]

We anticipate that the range of society's potential responses to environmental problems will be wide; different markets will respond according to their overall attitudes to a number of factors, including belief in the credibility of governmental action, level of social cohesion and the value placed on wider society beyond the family. No-one is equipped with a crystal ball, yet there is value in scenario development to separate out potential trends and review their progress, as well as using them to test the robustness of internal strategies and scenarios. Each way forward offers challenges as well as opportunities for business development.

4.6 Techniques for developing strategy options

Within the company there are skills and a knowledge base to be harnessed while generating strategy options. Given the complexity of the environmental arena, the first step may well be initiated by the planning function, as an attempt to raise the profile of the issues involved. True strategic change, however, can come about only when that knowledge has been shared and the widest possible group is charged with the task of developing a strategy for the organization.

After the appropriate awareness raising and development of knowledge, a structured group process of working through some of the issues raised is appropriate. Such a process might well include use of external facilitators employing one of the techniques listed below. Through working together on the options and testing out the conclusions reached, commitment to the joint goals of the organization is developed.

Analysing the present

CONTEXT ASSESSMENT

As with any traditional planning process, one needs a sense of what the outside world is doing and this requires thought on many of the issues raised in Chapters 1 and 2. In the light of the individual company's circumstances and environmental impacts a view needs to be formed of the environmental dimension of:

The operational environment: how this relates to internal processes.
Time horizons: when is change anticipated in your company, within the

other companies being used as benchmarks, within the industrial sector in which you operate, within consumers and customers and within wider society? Do you know the time spans involved in bringing about shifts of behaviour both externally and internally?

Product performance: does the current product structure or formulation work in the changing context of cradle-to-cradle management and post-use recycling systems.

Competitors' activity: have competitors been aware of and acting in response to these issues already? Do they have current products or services in the marketplace with environmental claims or benefits? Do they have technologies currently available, or in development, that have environmental benefits?

Strategic alliances and collaborative activity: the threats inherent in responding to environmental issues offer much room for joint action; investments in developing best practice, researching new technological solutions and providing information to stakeholders are all activities that can be shared in the right circumstances. Suppliers, retailers, distributors, franchisees, industry associations and trade bodies, pressure groups and charities as well as competitors all have contributions to make. Strategic alliances may crystallize from the general discussion of the issues involved.

Whole systems analysis: has the company compiled a metabolic chart of its environmental inputs and outputs and a stakeholder review? (See Chapter 5, Figure 5.3 and Table 5.7.)

Strategy delivery mechanism: does the environmental dimension alter capacity to implement a strategic decision?

Armed with a wider view of one's position it is possible to evaluate key determinants of a strategic shift.

ENVIRONMENTAL CHALLENGE FOCUS

Rather than attempting to create general scenarios of the future (as in Scenario Planning below), developing potential responses to a particular environmental issue that will impact on the industry sector involved, such as global warming, can be a useful focus for exploring the strategic horizon. Using creative problem-solving techniques to explore potential changes to the operating context allows a new perspective to be built and strategic options generated based on these drivers of change.

CONSENSUS BUILDING

The Natural Step process in Chapter 2 is an example of consensus building operating to improve environmental awareness. For the individual company, attempting to initiate such consensus-building exercises to find industry-wide agreement on technological innovation or demand-management strategies could be a way to avoid expensive investments which could potentially be bypassed by competitors and shifts in government policy. An industry with a clear agreed perspective on environmental matters, such as the industry-wide agreement reached by the European Chemical Industry to create an internal self-improvement programme, has the potential to work alongside government to mutual benefit.

By acknowledging the importance of environmental issues and coming up with positive options, an industry creates its own terms of reference, which in turn gives government a reference point for policy options and the means potentially to implement whatever overall targets are agreed. The example of the chemical industry's commitment to a well-financed voluntary 'Responsible Care' programme, although not 100 per cent successful so far, contrasts with that of the energy industry's approach to the issue of global warming, where no clear consensus has emerged on how reductions in global warming gases should be achieved, except defensively in opposition to carbon taxation.

BENCHMARKING

Like other areas of management thinking, environmental management is causing a rethink of working practices in strategy making, and companies interested in changes within the discipline are looking externally to others' experience for models to improve their own processes. Participation in benchmarking exercises is now coming to the notice of companies with sophisticated strategy and planning systems.

Predicting the future

DELPHI

There may be circumstances when the issues under consideration require an expert perspective, or the situation is so uncertain that there is a wide range of views as to the shape of the future, for example the development of the environmental performance of a particular sector. The Delphi technique was developed to manage collective inputs into a decision-making process during such periods of uncertainty, to use the range of views that are available to good effect by focusing positively on what discrepancies exist, to explore, through controlled feedback, a collective viewpoint. The method is specifically designed to correct the potentially false output of a face-to-face group struggling with internal politics, where minority opinions can so easily be suppressed or withheld.

The feedback process is conducted by the use of questionnaires, administered by a non-participating facilitator, and the responses are not attributed, thereby reducing pressure to conform to a dominant viewpoint. By conducting the analysis in a series of rounds, controlled feedback is given and continues until the picture stabilizes and a group perspective emerges. Although there is an expectation that feedback will lead to a consensus, this is not always the case in complex or genuinely uncertain circumstances, although the forecasts given will be more rounded than would be possible through other techniques.

ADVISORY PANEL OF EXPERTS

The use of outsiders to comment on and review strategy is not unusual; where innovation is taking place is in the use of stakeholder advocates as part of a strategy refinement process to generate and/or test options as shown by Example 4.2 below.

Novo Nordisk

In March 1993 Novo Nordisk invited 12 key European environmentalists to debate genetic engineering's potential for sustainable development. The meeting was the third in a series of open debates with agenda-setting 'green' activists to discuss fundamental aspects of Novo Nordisk's strategic development.

One of the main ideas behind the debate was to challenge the company's thinking: Novo Nordisk believes itself to be contributing to sustainable development and wanted to find out whether outsiders would share that view of its products and operations. Secondly, the meeting allowed issues around biotechnology to be discussed face to face (given the widespread concern about the technology within environmental groups), and allowed a continuing dialogue on the safety and environmental aspects of gene technology to be created.

Novo Nordisk believes that such exchanges have influenced its overall strategic response and led to a modified corporate communication policy, with more concern for stakeholders' interests.[14]

Example 4.2 Novo Nordisk

TRACKING SOCIAL ATTITUDES

Monitoring the speed of response to environmental issues is difficult without developing some awareness of the attitudinal changes taking place within wider society. Such change is particularly relevant to a strategic approach to the issue of needs and wants discussed in section 2.10.

Work has been going on since the 1970s to adapt Maslow's 'hierarchy of needs' to social behaviour, in order to determine how this is driven by individual motivations. The work of SRI International, an offshoot of the Stanford Research Institute in the USA, and Christine MacNulty of Applied Futures in the UK during the late 1980s, has created a highly effective model of the core motivational behaviours of the population of OECD countries. Their analysis centres around three core types – 'sustenance-driven', 'outer-directed' and 'inner-directed' – as three distinct groups within Western industrial society whose attitudes are sufficiently differentiated for different approaches to life to be evident. The emphasis is on attitudes to life rather than class, level of material wealth, age or gender.

'Sustenance-driven' people have a primary need for security, working in the here and now and belonging to a group of like-minded peers. Within this group there are those who place more emphasis on surviving from day to day and others who feel that 'belonging' is more important. On the other hand, 'outer-directed' people seek enhanced self-esteem externally through status and doing the right thing; this group varies in its attitudes to materialism and conservatism, but the general thrust is self-interest. Finally, there is the steadily growing 'inner-directed' group, who seek to succeed within their own terms of reference, in Maslow's terms to 'self-actualize', and who care about their ability to develop socially, emotionally and spiritually as well as being concerned about the quality of life offered and the state of the environment.[15]

Strategically it is important to be able to evaluate how society is changing by shifts in the overall focus of policy and values, and this model is very helpful in monitoring such subtle developments. The percentage of the population in these broad groups is changing over the decades, and the mix of such groups in any individual territory will have an impact upon its political, economic and social systems as well as on the economic choices of individuals. Monitoring these changes is an important dimension to the strategic planning processes and one that informs both the overall operating climate and potential consumer attitudes.

SCENARIO PLANNING

Some circumstances involve so many potential fluctuations within the operating environment that coherent pictures of possible alternative futures are desirable, and scenario planning can be very useful in these circumstances. Instead of forecasting in a linear model, at times when there has been much dislocation, scenarios are used to create a set of stories about alternative possible futures, thereby promoting discussion of possibilities other than the 'most likely' one and encouraging the consideration of 'what if' questioning.

The long-range investments involved in the energy industry led to the development of particular expertise in strategy scenarios. Shell International is a leader in scenario planning, and the following sustainability scenario was developed in the period 1988–90 with the input of outsiders, including professional environmentalists, to inform group-wide thinking.

Shell International

This scenario is covered in the Shell publication *Global Scenarios for the Energy Industry*[16] and represents an attempt to make sense of the turbulence being experienced at the end of the 1980s. On the basis of three areas of fundamental change – geopolitics, economics and environment – Shell has created two alternative scenarios to represent different outcomes from current trends: 'Global mercantilism' and 'Sustainable world'.

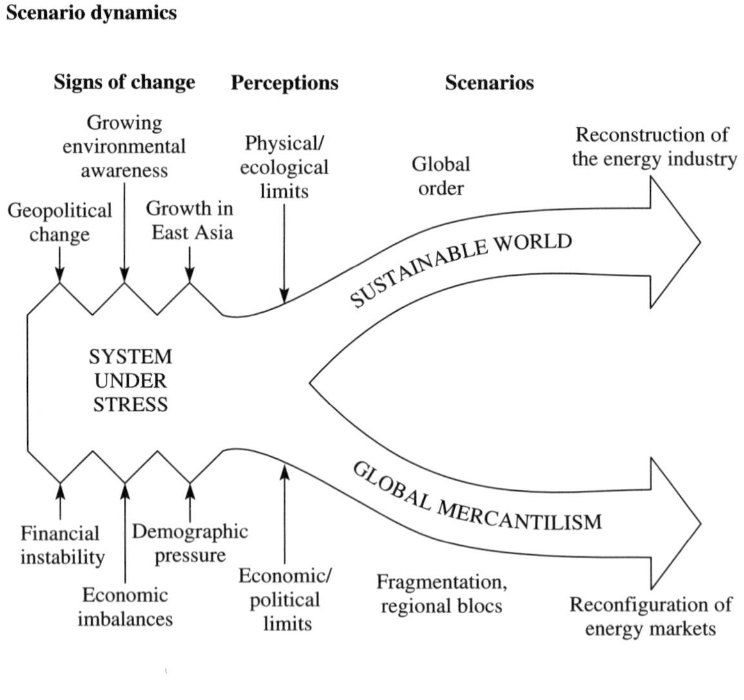

Figure 4.1 Two alternative interpretations of the present signs of change

The logic of the sustainable world

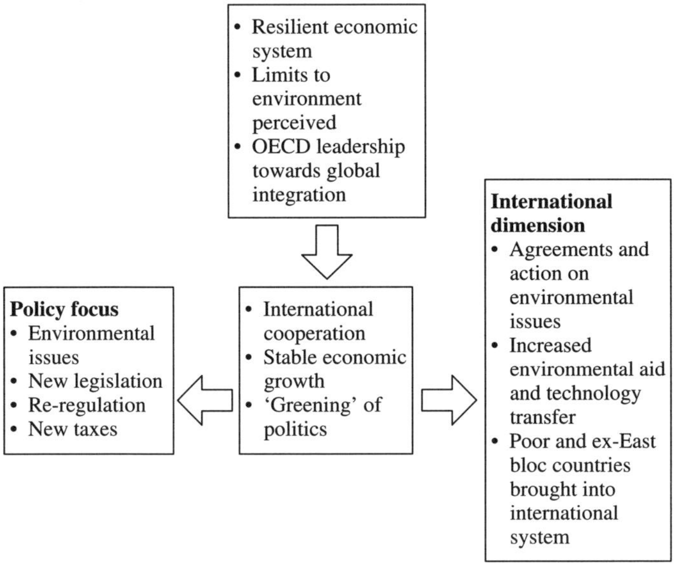

Figure 4.2 The logic of the sustainable world

The two alternatives represent very different logics and with them bring very different outcomes for the global energy market (see Table 4.1).

Table 4.1 Summary of the scenarios

	Global mercantilism	*Sustainable world*
Challenge	Hegemonic decline and economic instability	Degradation of the environment (especially through global warming)
Response	Multipolar world and mercantilism	International cooperation and management
Implications for energy	New rules for business and reconfiguration of markets	New values for fuels and reconstruction of the energy industry

4.7 The progression of strategic awareness

Research into the strategic response of organizations has shown that the development of an environmental dimension takes time to permeate through the organization, and a progression can be seen as the idea matures internally. The following model of corporate capability, based on the survey carried out by the Prospect Centre and the Environmental Management Centre at Brunel University,[17] is qualitative and fairly impressionistic. Each stage has two facets – first, the general strategic capability to develop and change, and secondly, the strategic capability to implement environmental choices.

Corporate capability to respond strategically to the environment

STAGE ONE: BEGINNER

- Environmental protection is not on the senior management agenda, or is seen only as a threat to business performance.
- The business has little sense of longer-term direction or competitive strategy, and few processes which could support the pursuit of direction and strategy.
- Regulators, and even customers and suppliers, tend to be viewed as adversaries.

STAGE TWO: EMERGENT

- The business has an environmental protection policy, and environmental protection is one factor in business decisions taken by senior managers.
- The business has explicit longer-term business goals and a number of strategies to support competitiveness, but no explicit environmental focus to these goals and strategies.
- A limited range of information is being systematically collected which could alert senior managers to threats and opportunities in the business environment.

STAGE THREE: DEVELOPING

- An environmental protection focus has been embedded into the longer-term aspirations of the company as a support to the main thrust of the business, and environmental protection is an important factor in senior management decisions and the development of management systems.
- Several of the company's strategies to support competitiveness have an explicit 'green' focus.

- Information is being systematically collected from environmental audits/reviews and environmental impact assessments, and a range of other sources of information about threats and opportunities in the business environment are being systematically used.
- Partnerships with stakeholders are being formed.

STAGE FOUR: MATURITY

- The business is breaking new ground in its innovation strategies and environmental management systems; environmental management is seen as a longer-term opportunity for business growth and is central to senior managers' thinking about the business.
- Strategies to support competitiveness are highly developed and have an explicitly environmental focus; responsibility for competitive strategies is placed corporately where it can give the greatest benefits to successful innovation.
- Information from a wide range of sources is used to develop competitive strategies, and there is a pronounced environmental dimension to the information that is collected and used.
- Relationships with stakeholders are being actively used to leverage innovation.

STAGE FIVE: SUSTAINABILITY FOCUS

- Sustainability is one of the core values and a key part of the vision of the business; it is central to innovation strategy and to senior managers' business decisions.
- All strategies to support competitiveness have a high-profile sustainability focus, including particularly finance and human resources.
- Information-collection processes for use in strategy development are the subject of innovation themselves, and are driven in part by the explicit need to define and realize sustainable competitive strategies.
- Complex and multiple alliances with a variety of stakeholders are being exploited.

There are many stages to be negotiated in the move towards a business strategy based on sustainability as a core value; however, we trust that the days of business strategists neglecting the environmental dimension of their operations are now numbered.

References and further reading

1. The Polluter Pays Principle, as enshrined in the UK's 1990 Environmental Protection Act, following its adoption in 1972 as a guide to environmental policies by the industrial members of the OECD.
2. S. Schmidheiny, *Changing Course*, MIT, 1992.
3. Exceptions to this general indifference include the growing number of 'ethical' funds in the US and Europe and the Cooperative Bank, which in the UK made its ethical lending policy the basis of a major advertising campaign to attract customers away from other high street banks.
4. Smythe, Doward and Reback, *Corporate Reputation: Managing the New Strategic Asset*, Century, 1993.
5. Ulrich Steger, *Umwelt Management* (in German), Frankfurter Allgemeine/Gabler, 1988.
6. John C. Newman and Kay M. Breeden, 'Managing in the Environmental Era', *Columbia Journal of World Business*, Vol. XXVII, Nos III and IV, 1992.
7. One example is The Body Shop, which is planning to balance its use of non-renewable fossil fuels by investing in an equivalent amount of wind-generating capacity.
8. Business in the Environment (BiE) with KPMG Peat Marwick, *A Measure of Commitment: Guidelines for Measuring Environmental Performance*, BiE, 1992.
9. BiE with the Chartered Institute of Purchasing & Supply and KPMG Peat Marwick, *Buying into Environment*, BiE, 1993.
10. Best Available Technology Not Entailing Excessive Cost – the principle enshrined in the pollution prevention sections of the UK Environmental Protection Act of 1990.
11. N. Fonda and K. Knight, *Prospering through Environmental Protection: The Strategic Management Development Challenge*, Environmental Management Centre, Brunel University, 1993.
12. Jeremy Leggett, 'Who will Underwrite the Hurricane?', *New Scientist*, 7 August 1993.
13. National Environmental Policy Plan 2, Ministry of Housing, Spatial Planning and Environment, The Hague, Netherlands, December 1993.
14. Sources: Novo Nordisk's own internal newsletter *Environmental Update*, together with interviews with senior members of the Environmental Services Department.
15. For a fuller explanation of this work, see Francis Kinsman, *Millennium: Towards Tomorrow's Society*, W. H. Allen, 1990.
16. A. Kahane, *Global Scenarios for the Energy Industry: Challenge and Response*, Shell International Petroleum Company Ltd., 1991.
17. N. Fonda and K. Knight, op. cit.

5 Financial strategy and environmental accounting

1. The environment is our most important capital asset, and this needs to be recognized in our financial strategy.
2. There are two aspects to integrating the environment into management accounting: extending the accounting systems to incorporate environmental costs and benefits; and applying financial models to the environmental impacts of the business.
3. Incorporating the environment into the normal management accounting and control systems involves changes in a wide range of existing systems, such as:
   - costing systems (including energy and waste accounting)
   - the planning and budgeting system
   - capital budgeting and investment appraisal
   - plant maintenance
   - performance measurement, appraisal and reward systems
   - forecasting
   - purchasing policy
   - research and development policy
   - business cases, business acquisition and development
   - control of subcontracting
   - insurance policies and environmental risks.
4. Impact and Resource Use Accounting, which applies financial models to the inputs and outputs exchanged between the company and the natural environment, is an important development area for financial management, in view of the likely demand to include remediation in all costs and prices.

5. There should be increasing involvement of financial accountants in external reporting on environmental performance. We outline the requirements and chart the stages of development towards a 'Sustainable Business Report'.
6. Environmental performance will also play an increasing role in the attitudes of investors, pension funds, banks and other creditors.
7. Financial planning of the capital base, including significant investment and divestment, and moves towards greater financial disclosure, will form an important part in any attempt to move towards sustainability.
8. Five alternative financial strategies, together with related issues and techniques, are charted under the headings of:
   Compliance-driven
   Resource use reduction
   Risk reduction
   Business opportunity
   Progress towards sustainability.

> Only after the last tree has been cut down, only after the last river has been poisoned, only after the last fish has been caught, only then will you find that money cannot be eaten. (Cree Indian prophecy.)

5.1 The ultimate capital resource – the environment

Any business which wishes to be successful distinguishes between its capital base and the income it generates. The responsibility of managers and owners is to husband both types of wealth carefully. As we learn more about the environmental challenge we face, wider society is beginning to apply these disciplines to global *ecological capital.*

Ecological capital is the basic environmental resource base on which human activity depends – the soil, forests, seas, lakes and rivers, mineral and other reserves which human beings exploit and the water, energy, fish stocks and fertility which we take from them. Like a business, human society depends on the use of *capital* (environmental resources), *investment* (savings being recycled to maintain productive capacity – fertility or stock levels) so that the earth can go on being available to produce the *income* (production of commodities and raw materials) on which we all live. Unlike a successful business much human activity takes income without paying the price of reduced consumption to set aside a portion to make good depreciation of capital, and in many instances we take resources which we are in no position to replace – the essence of an unsustainable practice.

Sustainable living means applying the sensible discipline of capital maintenance to all ecological capital, thereby actively protecting our capacity to enjoy long-term income and maintain humanity's standard of living. At the individual company level, implementing sustainable operations begins with a strategic review of the financial strategy and systems, ecological capital costs and how to incorporate the new environmental dimension of business.

5.2 Integrating the environment into financial strategy

We believe it is crucial for the long-term development of companies themselves, as well as the emerging discipline of Ecological Management as we have outlined it, that environmental concerns are integrated into financial decision making. Financial directors and managers, who play a key role overall as business managers, need to understand the contribution their discipline can make to improving environmental performance and moving the enterprise nearer to sustainability.

Traditionally business has been reluctant to deal with the wide range of issues raised by environmental concern, in part because of their multidisciplinary, non-linear nature. Unlike human-created resources, however, much of the capital base that nature has created cannot be re-formed within a human time-frame; it cannot be discounted or written off in the books. Without the continuation of natural systems, like the water and carbon cycles, life for human beings will not be possible. From that perspective the need to account for ecological capital just as rigorously as accountants track and control 'man-made' capital is obvious. This would be an extension of financial analysis and management, applying its rigour and techniques to managing the way in which organizations impact on, and utilize, the natural environment.

Viewing environmental issues from a financial perspective does, however, require the development of new knowledge and a change of attitude. Yet those trained in the methods of accountancy and corporate finance will be pleased to know that there are similarities between ecological management and typical exchequer concerns about the allocation of resources, or the analysis skills needed for portfolio management and financial modelling for a whole organization. An enterprise seriously in pursuit of sustainability needs auditing and project appraisal skills, both to understand and quantify the damage incurred by its current practices, as well as determining the costs and capital transfers involved in moving to more sustainable means of making a living.

This is a very new approach, and there are few benchmarks for individual companies to emulate. As well as sharing information with other pioneers, it will be necessary to work from first principles. Perhaps the best place to start is to differentiate between the extension of existing management accounting systems to integrate environmental issues into the workings of the company and the use of financial models and skills to track the environmental impact of the business. Both challenges necessitate a change of perspective and the development of new systems.

Much management or cost accounting for the environment is organized by the environmental team, with the limited assistance of financial personnel within the organization. Currently, accounting for the environmental *impacts* of the organization is often the exclusive preserve of environmental management. It is in this latter area of accounting for the environment that we believe financial skills should be applied, both to improve the practice and to bring to bear accountants' better understanding of ecological capital accumulation and maintenance.

The current separation of functions represents the mental separation evident in many organizations; the money model of operating marginally adapted to count environmental costs, while the environmental impacts of the organization are separate and lodged beyond the accountants' influence in the environmental manager's domain. Until this gap is bridged there will be strict limits on what can be achieved, as the financial controls of many organizations are set up in such a way that environmental improvement is severely hampered, being blocked at many stages by the constraints built into the financial system.

5.3 The environment and management accounting and control

We believe it is fundamental to ecological management in the longer term that environmental considerations are integrated into the normal management accounting and control systems of the organization. We are under no illusions as to how extensive a change in practice that would entail. Professor Rob Gray and his colleagues Jan Bebbington and Diane Walters from the University of Dundee have published, in association with the UK Chartered Association of Certified Accountants, one of the first works on this subject *Accounting for the Environment*.[1] Table 5.1 indicates how far, in their view, these changes must reach.

Environmental data to support decision making can be more easily generated in a company which has a clear corporate environmental strategy and in turn has reviewed the full financial implications of the environmental agenda. Let us look in detail at some of the areas where environmental concerns will impinge within cost and management accounting.

Table 5.1 Environment and management accounting and control issues

Costing systems
Planning system
Budgeting system
Variance analysis
Capital budgeting and expenditure
Plant maintenance
Investment appraisal and post-audit
Performance measurement and appraisal
Reporting and accountability systems
Reward systems
Forecasting
Scenario assessment
Options review
Purchasing policy
Research and development
Business cases
Business acquisition and development
Control of subcontracting
Risk assessment
Insurance policy
Financing decisions

Source: Rob Gray *et al*, *Accounting for the Environment*, © Paul Chapman Publishing Ltd., London, 1993, page 149.

The costing systems

The relationship between conventional accounting systems and the environment is such that a number of very basic questions need to be addressed. Does the system even identify what costs are being incurred as environmentally related expenditure, both overall and the environmental aspect of new processes? Can the organization identify the actual costs of the environmental resources it uses, such as energy or water, or the level of dependence on existing environmental sinks through waste disposal to landfill or sewerage discharge?

The above may seem more obvious than the costs of meeting external environmental criteria in the form of consent and compliance costs to meet legislation, such as the UK's Environmental Protection Act EPA and Integrated Pollution Control IPC regulations. It is worth bearing in mind the management time spent on environmentally related issues, including responding to crisis situations.

When evaluating these figures, have systems been established to account for both the expenditure *and* the savings made, even if the savings accrue put into estimating the financial dimension of that policy decision. Many of the changes identified in this book will have financial ramifications, and not all of them negative. The costs involved in remediation of current non-sustainable practices will need to be gathered and the introduction of remediation planned to ensure the ongoing viability of the enterprise, as it relinquishes use of ecological margins.

The budgeting system

If the environmental strategy of the company is to be implemented, this will require local managers to have the freedom to make the best choices possible within the overall strategic mission. Does the system currently have environmental budgets and targets for individual companies, divisions or activities? If there is a company-wide commitment to BS 7750 or the EU Eco-Audit and Management Scheme (EMAS), have budgets been set aside to allow for its introduction, with adequate training and awareness generation, or have individual budgets been increased to accommodate the extra resources required?

Indeed, some companies have decided against developing a special environmental budget, fearing that financial demarcation would lead to a narrow focus of responsibility for environmental protection, as the province of environmental management. Instead of a separate cost centre they have decided that environmental expenditure should be included in departmental budgets. Major expenditure such as environmental audits and reviews, however, especially those involving external consultants, do involve resources. Somewhere in the organization someone must be able to allocate such costs to their own budget, if a central environment budget is to be avoided. Companies with quality-management-based systems may find a non-centralized budget structure easier to achieve than those without this existing infrastructure.

Capital budgeting and investment appraisal

For many sectors, improved environmental performance is dependent upon investment in cleaner process technologies, and this will necessitate capital expenditure and adequate planning for the phase-out of old processes and the introduction of new systems. In other industries, such as those dependent until recently upon CFCs, the whole nature of their technology base has been redefined by environmental concerns, and this has played a major role in the financial health of the operations involved. In other sectors, their ordinary activities and investments have substantial environmental impacts which need to be considered. Elsewhere, planning for environmental improvements is appraised on a case by case basis, and the hurdles which environmental projects are set, as against other types of investment, need to be considered.

Each of these different circumstances requires new financial appraisal policies if the business environmental strategy is not to be stillborn. Looking at many of these changes solely against financial criteria distorts the overall perspective until change begins to look as if it is not affordable. Traditional investment appraisal techniques such as 'Payback', 'Discounted Cash Flow' and 'Contribution to Profit or Earnings per Share' may not be particularly appropriate to environmental issues. The exclusively financial nature of these systems with their bias towards short-term and lesser risk options will clearly be a disadvantage to environmentally-driven projects of a long-term return or those set up to anticipate prospective changes to operating conditions.

Responding to ecologically-driven changes to the operating environment is to some extent like responding to any other external market signal. Unfortunately, the lack of unanimity about the scale and timing of the changes required means there will be wide gaps between the winners and the downright losers. Reappraisal of current business products or services and the processes that bring them to market is a key background to making investment decisions. In the face of this uncertainty and the need to make markets by innovation and new practices rather than be passive, companies with strong environmental policy commitments may choose to bypass their traditional evaluation processes on the grounds of policy. Table 5.2 represents a checklist for realigning appraisal systems to consider this wider dimension.

Table 5.2 Checklist for developing environmentally sensitive investment appraisal[3]

1. *Environmentally screen all investments*

All investment proposals should be accompanied by an explanation of their environmental impact or case, which for major activities covered by legislation may involve a formal 'Environmental Impact Assessment'. For lesser projects a mini-version would be appropriate.

For those unconvinced of the need for more radical strategies, compliance with legislation necessitates understanding and including in your investment criteria, then achievement of, 'Best Practicable Environmental Option' (BPEO) and 'Best Available Technology Not Entailing Excessive Cost' (BATNEEC). Companies operating prescribed processes under the UK EPA are legally required to screen investments to ensure that they comply with these constantly upgrading objectives, and, should there be an investigation into a company for breach of consents, having an audit trail to show evidence of this process of evaluation will be necessary.

2. *Reconsider costs*

The anticipated benefits of conventional proposals may disregard the environmental costs, both financial and non-financial, of the planned activity.

3. *Reconsider benefits*
When presented with environmentally linked investments have all the paybacks been presented as well as the costs? Many of these investment opportunities do have financial payoffs if these are considered, even under conventional assessment criteria. Environmental proposals may underplay their benefits in waste reduction, avoidance of anticipated cost increases and minimization of compliance costs and management investment.

4. *Reconsider the criteria applied*
The need to marry the environmental objectives of any organization with its investment criteria is paramount. Unless the environmental dimension is formally acknowledged by the board of the company and incorporated in the criteria for financial appraisal, much innovation will be blocked for lack of financial support.

5. *Reconsider the possible options considered*
Looked at from the perspective of the whole organization, does this proposal simply provide a solution in isolation, with savings being overlooked? Would an alternative solution be both environmentally superior and give a better return? For example, does the proposed investment involve a bolt-on clean-up to reduce environmental damage, when this might be better achieved by re-engineering the process to avoid making the waste in the first place? Or could improved efficiency be achieved by utilizing waste materials as fuel instead of investment in more energy control mechanisms?

6. *Consider the opportunity costs*
Has there been a serious analysis of the costs of not accepting a proposal; for example, to set in train recycling systems for the organization, against the rising costs of waste management based on disposal?

7. *Reconsider the time horizon*
The results of environmental investment can be speedy, for example in changes of process to eliminate waste, but more realistically paybacks may not occur within the narrow time-frames laid down by current policy, e.g. 12 months.

8. *Reconsider the discount rate*
Existing discount rates frequently make long-term investments unviable, particularly in environmentally benign technologies. Discounting also takes no account of the full costs of remediation to be undertaken by the future company or the society of which it is a part.

Making adjustments to take more account of the medium term (3–5 years) is a beginning, but leaves one distanced from the concepts behind sustainability of remediation and ecological capital balance which may take decades to achieve. There are arguments that environmental projects should use a lower discount rate or, indeed, a negative rate to reflect their environmental dimension.

9. *Consider the valuation of 'externalities'*
Would your investment appraisal decisions be rather different if environmental resources, such as water, air, waste disposal to land or water, were priced at their true environmental costs, that is, including a remediation contribution? Will the

anticipated changes in their costs over the coming decades change the costs or benefits involved in the investment?

10. *Consider decisions in the light of sustainability*
Viewing investment decisions of acquisition, disposal, market development and process re-engineering from the perspective of a commitment to sustainability will inevitably change reactions to investment proposals. Are the plans on the table one step in an overall strategy of improvement, or could one leap-frog several steps by looking at the issues from a different perspective?

One example of changed investment appraisal to incorporate environmental criteria is that of the Fiat Auto group, one of Europe's leading industrial companies. In this organization all sectors review capital expenditure planning and work programmes as part of a coordinated approach to environmental problems from a long-term perspective, including:

- intelligent product research, testing and design
- optimization of all manufacturing processes
- careful assessment of new technologies
- proper waste disposal
- research into urban traffic systems and more rational ways of integrating public and private transport.[4]

Systems to support major capital expenditure need to change substantially to incorporate environmental performance and sustainability objectives.

Plant maintenance

Organizations which have piloted best practice have found that achieving operational improvements necessitates a new emphasis on maintenance, both to minimize hazards and accidents and also to ensure that energy and other resource use is reduced by maintaining maximum operational efficiency. One example would be the replacement of bulbs with energy-efficient models and cleaning of light fittings to maintain light levels, increasing unit output efficiency and thereby reducing energy costs. Although there is much documentary evidence and research to support this incremental way of operating, unless internal accountants uphold the approach and are prepared to look beyond the short-term opportunity to cut maintenance budgets savings will be lost and impacts maintained at a higher level unnecessarily.

Performance measurement and appraisal

The 'performance measurement' and 'appraisal' systems that make the company function effectively must include environmental achievements in an organization that is committed to improved performance. The old adage that what is measured is what can be controlled works as well for environmental matters as any other area of activity. Currently, few operations include environmental indicators in their financial performance targets, and much work is still to be done to create an index for firms to emulate.

Reward systems

A company which is truly serious about its environmental status will be working to incorporate environmental performance achievement into the managerial appraisals of members of the company. It is critical that under-performance in environmental matters is noted within the reward system, for without a challenge to those prepared to pay lip service or block environmental change realigning attitudes will be extremely difficult.

Gray quotes the example of a company operating within a Total Quality Environmental Management (TQEM) culture which 'expects all target areas of the managerial appraisal system to be met within a framework of zero complaints (internal and external), zero notifications from the regulatory authorities, zero transgressions of consents or internal standards (whether notified or not) and a rapid recognition of suggestions from external or internal sources'. Both financial remuneration and internal promotion are dependent in this company on achievement within this framework; here, clearly, the environment is not a token activity.[5]

Forecasting

The anticipated changes to the global operating environment outlined earlier in Chapters 1 and 2 do, we believe, necessitate a fundamental review of the medium- to long-term financial projections of any company.

Significant impacts on the cost base include anticipated increases in energy costs and waste management, increased insurance premiums to cover environmentally sensitive sectors and the additional management time and expenditure involved in environmental compliance. This, coupled with the longer-term implications of contaminated land liabilities and the bearing this may have on asset values, together with contingent liabilities that would result if the insurance industry withdrew from more aspects of environmental cover, suggests that the financial director needs to understand the risks and opportunities associated with environmental issues.

Does the company's financial forecasting process review these questions? Is the company planning for the changes in environmental costs anticipated? Is the future of the existing technology base of the organization secure in a world where the polluter is required to pay for any environmental impacts? Are changes likely in raw material prices as these rise to incorporate remediation charges, and will these rises reduce margins? Alternatively, will market expectations for environmental quality lead to a loss of margin for companies as they are forced to make the required improvements? Will lack of environmental quality undermine the overall perceptions built up around the brand or company over time? What would be the financial dimension of a major pollution incident or court order for damages?

Scenario assessment

Is there an environmental dimension to your scenario planning? Is the company able to envisage the financial changes which would be brought about by a move towards sustainability within society at large? Does the company have survival plans should the costs of major environmental resources change, such as a major hike in energy costs, due to the imposition of carbon taxes, or the withdrawal of licences to sell particularly damaging products, such as the chemical industry has had to deal with?

Purchasing policy

For the company with a comprehensive environmental policy, sooner or later a judgement will need to be made upon the application of environmental criteria to purchasing. This review leads inevitably to consideration of the financial implications of positive discrimination in favour of environmentally benign products and suppliers, as many of these goods and services cost more, unit for unit. The additional expenses to be incurred can be offset by savings in waste elsewhere by other environmental policies. When additional costs are incurred, can these be subsidized by other parts of the organization's environmental budgets? (See Chapter 9 for a fuller discussion.)

Research and development (R and D)

Financial criteria for R and D expenditure that is focused towards environmental problems and business opportunities may, like other environmental investments, require a different assessment basis. This would have to take into account the potential longer-term penalty of failing to address the problem early and being forced to pay royalties to use solutions developed by others or leave a business sector completely. Adaptations to product and process design changes should not be blocked by short-term financial criteria, if these will be vital to product performance or to meet regulatory standards in the years to come. Organizations such as

Alcan and ICL have included environmental criteria into their design appraisal processes, believing as they do that the environmental performance of their product is critical to its competitive performance. (See Chapter 7 for a fuller discussion of these issues.)

Business cases

The financial performance of an activity, product or organization can be severely affected by underestimating the environmental aspect of operations. Business cases built on environmentally adverse operations, which bring with them contingent risks, or inappropriate cost projections which ignore substantial costs, such as the need to upgrade technology and processes to meet higher standards, may not be commercially sound. Much of the City of London's infatuation with waste management as a 'green business' sector of the equity market during the 1980s has gone sour as the costs of remediation and compliance control of existing facilities have turned out to be higher than anticipated and landfill taxation threatens to further distort growth projections from 1996 onwards.

Business acquisition and development

Awareness is increasing of the financial risk inherent in buying into an organization without understanding its environmental liabilities. Many merchant banks advising corporate investors will now ensure that due diligence includes an environmental review for organizations in sensitive areas or holding large land portfolios. This is especially true in the USA due to the potential risks of land contamination and clean-up costs. What is not done as far as we are aware is an appraisal of the organization under consideration in terms of the environmental dimension of the business's long-term profitability. Would knowing more about the environmental profile of the organization undermine its position as a going concern or show increased risks?

Control of subcontracting

Corporate environmental performance can be undermined by the actions of subcontractors and long-term associates. Where agents are acting on one's behalf there is no defence in law should their failures result in environmental damage. In addition, where subcontractors manage non-core activities on behalf of the company, such as haulage, the contractors' willingness to improve their own environmental performance will limit customer's overall environmental performance.

Insurance policies and environmental risks

The insurance industry is in the business of managing risk and assessing the scale of potential claims against premiums received. Blanket environmental insurance is getting harder to come by as companies limit the ability to make environmental claims for accidental damage under public liability policies. Insurance cover is being split to distinguish between the nature of the incidents involved. 'Sudden and accidental' cover is for the one-off, uncontrollable incident, and different insurers will offer different cover for this type of event. The other type of environmental accident would be 'gradual pollution', and this is specifically being excluded from public liability cover.

Cover for such long-term damage as that involved in the now famous British jurisdiction *Cambridge Water Company* v *Eastern Counties Leather plc* case, where a tanning company unwittingly polluted aquifers with solvents in use in the post-war period and was sued for damages by the water company that lost use of an adjacent bore-hole, is now impossible to gain under public liability insurance. In that case £1.6 million damages were awarded in favour of the Cambridge Water Company, only to be reversed by the subsequent House of Lords judgment in favour of Eastern Counties Leather in December 1993.

The ramifications of that case for the insurance companies involved would have set a very significant precedent had damages been upheld. At the time of writing, the case is still being fought to determine the role of insurers in meeting anticipated clean-up costs. Insurers have a lot to lose as the growing number of environmental and health-related claims from old policies is causing major financial headaches for underwriters, as claims for compensation for asbestosis, other industrial-related diseases and long-term pollution have to be met.

A limited number of new plans, known as 'environmental impairment liability' (EIL) policies, have been created in the EU and USA to pay out when regulators order companies to clean up damage from gradual or accidental pollution and contamination. These policies are on a 'claims made' basis and the excess payable is often substantial. Many EIL policies specifically exclude coverage for pollution liability resulting from non-compliance with environmental legislation.[6]

It is essential for any business operating in sensitive areas to review its insurance cover and take the necessary steps to ensure that ability to cover potential liabilities is maintained. The alternative in many cases would be insolvency should a major incident happen and the organization be left unprotected.

The global insurance industry is also beginning to take the threat of global warming more seriously in the light of the high incidence of losses due

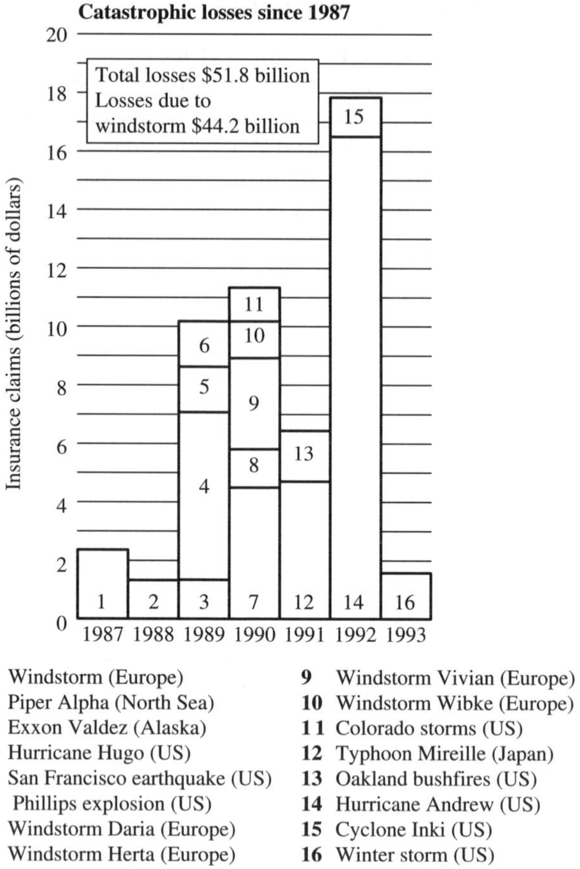

Figure 5.1
Source: J. Leggett, 'Who will Underwrite the Hurricane?' from *New Scientist*, August 1993, reproduced by permission.[7]

to environmentally related disasters since 1987.

The losses highlighted in Figure 5.1 do not take into account subsequent damage to property caused by landslides in California, bush fires in Australia and California and the disastrous flooding of the Mississippi since that chart was compiled. The losses incurred through environmentally related catastrophe insurance are said to be on such a scale that some analysts believe that the global reinsurance market may not be able to continue.

In some areas of the world such as the Caribbean, insurance is being withdrawn in the light of persistent storm damage. The scientific community is currently split on the subject of whether much of the freak weather being experienced is a result of global warming. When the cause and effect is established beyond doubt, companies throughout the world will notice reductions in cover against natural disasters such as flooding, storms, etc. as annual pay-outs outstrip premiums paid to insurance companies.

One of the most significant pieces of EU legislation for the insurance industry is the proposed Directive on Civil Liability for Damage Caused by Waste and the potential it gives for common interest groups such as pressure groups to initiate law suits. This is fuelling concern about environmental liability with general public liability insurance and causing reductions in the cover given. As this is a complex and evolving area, detailed advice should be sought from your own insurers.

For the individual organization seeking insurance there can be no doubt that insurance costs are set to increase, and this should be accounted for in future financial calculations. In the long term an insurance industry-sponsored system for assessing environmental good housekeeping by means of 'pollution ratings' may evolve, leading to premium bonuses for good practice, thereby making insurance more affordable. In the meantime companies can only try to negotiate with their insurers and use environmental improvements as demonstrable efforts to reduce risk.

5.4 Impact and resource use accounting

Somewhere between management accounting and financial accounting falls the whole issue of accounting for the environmental resources and impacts of the company. Currently, gathering information about the environmental performance of the organization is done in a variety of ways. The umbrella term for this might loosely be 'environmental auditing', although there are a great many types of data-gathering exercises, depending on who is collecting the information – from Environmental Surveys or Reviews at the start of awareness generation, to project or investment Environmental Impact Assessments, Eco-balance audits where the company reviews its overall process, to Life-Cycle Analysis where a particular product is reviewed from initial inputs to final outputs. Each technique depends on data capture.

Rigorous information gathering of this nature is slowly building a model of the full environmental impacts and resource use of organizations, and this will allow performance measures to be developed as the base information from which to develop sustainability and remediation strategies. To develop this knowledge necessitates creating new audit trails and finding methods of preparing and presenting data, and there are no standards to date. The following represents an honest attempt to explain generalized principles, for adaptation to your own circumstances. Each company contributing to the process of development is grappling with the issues from its own perspective, and many different methods are being used as organizations experiment.

Impact and resource use accounting is the monitoring of flows of ecological capital across the borders or 'cell walls' of the company, for example in the use of renewable and non-renewable material *inputs*, and outputs in the form of products as well as pollution and waste. Table 5.3 gives a generalized tabulation of some of the areas of resource use and pollution outputs made by the organization. This can be used as the basis of a system for quantifying some of the flows found in the corporate metabolism, as shown in Figure 5.2.

Table 5.3 Impact and resource use accounting

Natural Resources	*Energy Use*	*Emissions/ Effluent*	*Wastes*
Percentage of renewable stocks	Electricity	Emissions to air:	Solid wastes:
		CO_2	Hazardous
	Gas	NO_x	Non-hazardous
Percentage of non-renewable stocks		SO_2	Disposal to land on-site
	Diesel	Hydrocarbons	
		Methane	Disposal to land off-site
	Petrol	Particulates to air	
Recycled materials		CFCs/HCFCs/ Halon/Carbon tetrachloride/ 1,1,1 – Trich./ VOCs	Disposal to landfill
	Coal and solid fuel		Disposal to recycling contractor
Water consumption:	Air miles travelled		
From mains			Packaging by material type
Recycled	Renewable energy sources	Other	Oil recovery
Land use:		Waste water/ emissions to sewer:	Internal recyling
Contamination monitoring results		Biological Oxygen Demand	
New sites:		Chemical Oxygen Demand	
Reclaimed land		Suspended solids	
Land set aside for conservation purposes		Dissolved solids	
		Total organic carbon	
Contaminated land restored		Disposal to sea/river	
		Other	
Plant decommissioning and clean-up			
Site surveys (related to purchase and disposal)			

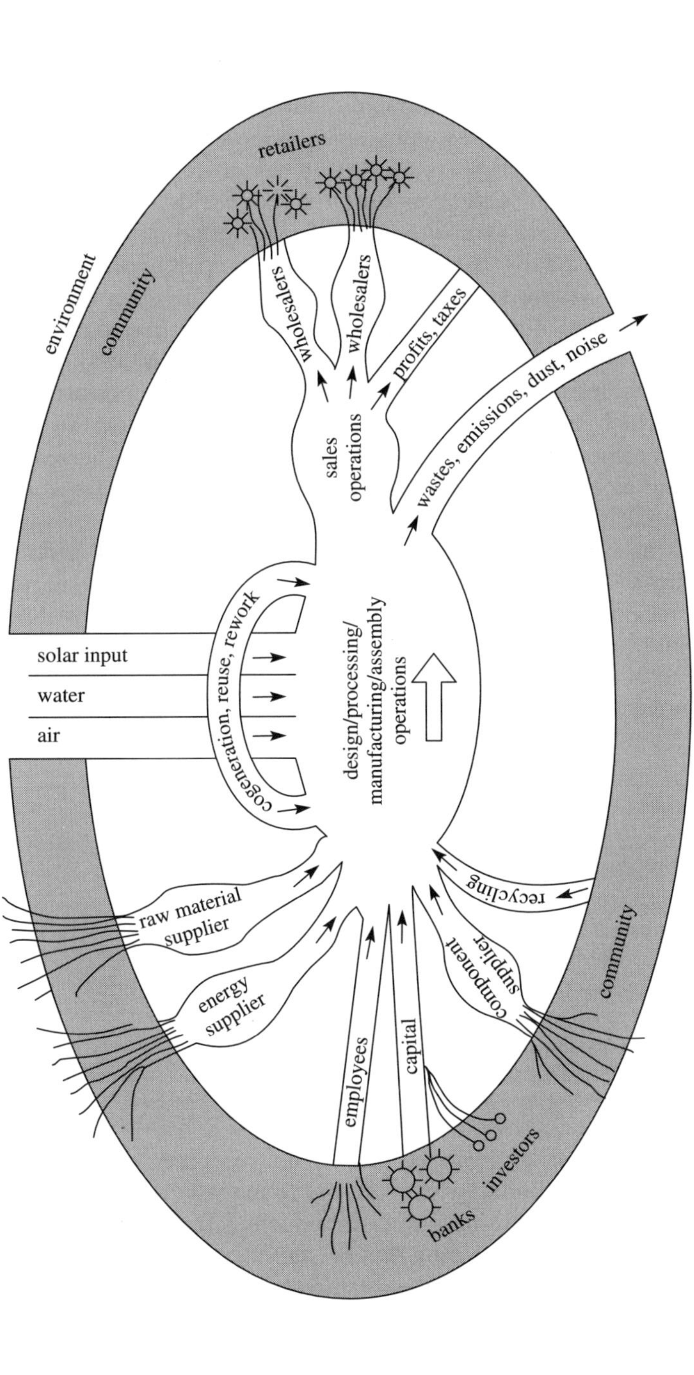

Figure 5.2 Simplified metabolic chart of a prototypical company
Source: E. Callenbach, F. Capra, L. Goldman, R. Lutz, and S. Marburg, *The Elmwood Guide to Ecological Auditing and Sustainable Business*, Berrett-Kechler Publishers, San Francisco, 1993. Reproduced by permission.

This type of information is being used in different ways. Its main purpose is to inform internal processes, to generate year-on-year improvement and to give a general picture of the environmental impact of the organization as a whole, as well as individual sites, companies or processes. Secondly, the data are collected for formal presentation as a rough picture of environmental performance of the company for the outside world.

Currently, how the data are presented for external comment and comparison is determined by the commissioning company; there are no formal measurement systems to allow real comparisons between organizations. The performance of two chemical companies offering data is of limited use as a comparison without such a standard. This weakness in existing practice is leading to the development of linked performance measurements, which attempt to place the data in a context so that benchmarking becomes possible and one is able to measure achievement over time. It is anticipated that a company recycling more per employee, etc. might be compared favourably with one recycling less and so on.

The measures in use by leading companies (shown in Table 5.4) attempt to put these flows into their operational context, although most of this activity is only meaningful externally when comparing like with like within a very narrow band.

Table 5.4 Performance measures

Performance measures
Cost/activity per unit of production
Unit generated per employee
Cost/activity as a percentage of total cost
Consumption/cost per m^2 (building use)
Tonnes of oil equivalent used per period, etc. (travel)
CO_2 equivalent per period, etc. (energy)
Expenditure as a percentage of total revenue costs (remediation)
Recycled as a percentage of total materials used
Recycled materials per unit of production
Number of miles/kilometres travelled

Understanding the range of impacts made and resources used by the organization is just the beginning when it comes to managing and operating sustainably. The sustainable business aims, in the long term, to be in balance in its use of ecological capital, ensuring that the capital consumed is replaced elsewhere. This can be achieved by investment in new technologies or through active regeneration of ecological capital stocks. This is the practice of active *remediation*, finding ways to replace the damage done, and is core to the aims of true sustainability. Only by applying remediation principles can we ensure that an alternative or the same type of ecological capacity is available to future generations.[8]

The principle of remediation is currently a totally alien concept to many in business, yet elements of this thinking are to be found in the Environmental Restoration Principle of the CERES Principles: 'We will promptly and responsibly correct conditions we have caused that endanger health, safety, or the environment. To the extent feasible, we will redress injuries we have caused to persons or damage we have caused to the environment and will restore the environment.' That commitment was developed as one of a set of criteria for ethical investment in response to the distress and grief occasioned by the Exxon Valdez disaster. Learning how to apply that sentiment throughout business operations is going to be a test of our ingenuity both as managers and residents of the planet. The full implementation of remediation as a principle will have major cost implications for future business activity.

5.5 Financial accounting and the environment

For many of the organizations pioneering work in this area it was the need to respond to external stakeholder concerns that prompted their own action. A key figure in handling external interest in the future will be the financial accountant, as public performance reporting involves wider concerns than those of public relations and may exceed the scope of the environmental team. We would also argue that financial strategy must be linked to environmental policy, with all the need for information, analysis and long-term planning that that implies. Both communicating the environmentally related dimension of company activities and ensuring that the company moves towards sustainability will increasingly involve accountants.

External reporting

As environmental reporting has matured from the basic bland statement in the traditional report and accounts, it has tended to move into the domain of the environmental manager. As information becomes more detailed and as the concepts behind sustainability are seen as being more aligned with traditional accounting concerns such as capital management, we believe the accounting function will be increasingly involved. This is why we have included external reporting in the financial strategy section of the book rather than with communication.

Disclosure to the public is very important to the basic credibility of business's ability to deal with the environmental question, so external environmental charters tend to make it a formal requirement. The ICC

Business Charter for Sustainable Development and the CERES Principles both require disclosure.

MINIMUM REPORTING REQUIREMENTS

The UK appears to be leading the field in debating the role of accountants in the provision of environmental information. The Institute of Chartered Accountants in England and Wales has recommended that a company publish the following details as a minimum part of the reporting cycle:

- the company's environmental policy
- the identity of the director with overall responsibility for environmental issues
- the company's environmental objectives which should be expressed in a way that enables performance against them to be measured
- information on actions taken, including details of the nature and amount of expenditure incurred, in pursuit of the identified environmental objectives
- the key impacts of the business on the environment and, if practicable, related measures of environmental performance
- the extent of compliance with regulations and any industry guidelines including, if applicable, whether the company's sites are registered under the EU EMAS scheme and the details relating to applications and approvals for registration under British Standard 7750 Environmental Management Systems
- significant environmental risks not required to be disclosed as contingent liabilities
- key features of external audit reports on the enterprise's environmental activities, including those relating to particular sites.[9]

They state that, where any of this information is provided in a document separate from the company's annual report and accounts, the latter should contain a reference to the availability of that document.

DISCLOSURE REGULATION AND STANDARDS

As the issue of quantifying and reporting the environmental dimension of one's business for external consumption is new to us all, it is full of uncertainty. There are no internationally accepted standards of regulation, and few countries, with the exception of Norway[10] and the Netherlands,[11] have measures to ensure that comprehensive environmental statements are published. There are some specific measures to ensure that certain types of environmental information reach the public domain, but these are localized in their effect.

The US Toxic Release Inventory, required under the 1986 Emergency Planning and Community Right-to-Know Act, requires companies with more than 10 employees to provide annual emission data for over 300 prescribed chemicals and a further 20 chemical categories when used in excess of set levels. In addition, the Pollution Prevention Act 1990 went on to extend reporting to chemicals recycled either on- or off-site, as well as waste reduction initiatives. This has created a huge amount of information available for use by outsiders, including the pressure groups. It should be noted that multinationals with commitments to consistent environmental performance criteria internationally may provide similar information in other markets.

In the EU such quality data are more difficult to come by, but the EC Directive 90/313 on freedom of access to information on the environment provides access to the wider public of all information in the public domain. Here the onus is on local authorities and regulators to release information to the public. This availability is a much more limited incentive for companies to disclose voluntarily through their own published public reports than the US model. The EU Eco-Management and Audit Scheme (EMAS) is also a voluntary scheme for manufacturing companies, but for those registered it contains mandatory disclosure commitments as part of the accreditation process. Some commentators are anticipating that the EMAS may well be extended to other sectors and be made mandatory once the initial trial period is evaluated.

The debate about regulating disclosure continues, due to concern over commercial confidentiality and potential litigation. Indeed, concern has even been expressed that disclosure of information voluntarily can be damaging to the company, and some lawyers are recommending constructing legal protections against potential adversaries.[12] Legal liabilities can result from ignorance as well as through the honest attempt at a response to environmental issues, and in recent cases of pollution litigation evidence of environmental management systems has counted positively in the favour of those who have inadvertently fallen foul of the law. Disclosure also has operational implications, as we discuss in Chapter 8.

The scale of the financial implications of certain types of environmental liability is so significant for operating performance that environmental reporting could be seen as a constructive process of identifying the true worth of an organization. The costs of clean-up of contaminated land are long-term liabilities which many companies do not currently declare. In ICI's February 1993 results, a provision of £148 million for remediation of agro-chemical sites in the USA caused surprise among the investment community. If ICI was not known as an organization with a comprehensive response to environmental matters, surprise may well have become concern.

The Confederation of British Industry has argued that 'Environmental reporting is a demonstration of serious commitment to a corporate environmental programme. Regrettably, it is proving to be the Achilles heel of the adoption of a full and open approach by business. Public reporting is absolutely essential to the credibility of any business, but some firms are thinking twice before committing themselves.'[13] The inference can be drawn that voluntary measures would be superseded by mandatory ones were the situation to continue.

Although not strictly public disclosure, providing information to those beyond the company is being forced by other pressures. The British Standard 7750, though a formal system of accreditation, has no targets included for performance, so companies wishing to set environmental standards for their suppliers cannot solely use the standard to ensure their aims, thereby leading the way to the increasing use of supplier auditing and back-up questionnaires during tendering. This may well increase the pace of external disclosure.

Avoiding reporting does not remove the external pressure for transparency in business practices; indeed the companies that pioneer reporting and performance measurement have a built-in advantage in being able to influence the debate on indicators which is ongoing.[14]

ENVIRONMENTAL REPORTING: THE RATCHET EFFECT

Reporting styles themselves follow a progression as the sophistication of the operation grows, expressing greater commitment to meeting stakeholders' information needs (see Table 5.5). The level of information available increases, together with the time, effort and trust invested. Courage is needed to overcome internal reluctance to improve the standards of environmental information given to the outside world. Experience seems to show that companies who invest in this area reap rewards in public relations, stakeholder relationships, employee confidence, improved ability to recruit staff and improved internal standard environmental management.

As far as we are aware to date, no company has yet completed the progression to full annual social and environmental sustainability assessment, but the Dutch company BSO/Origin has begun an interesting experiment which goes some way towards full cost accounting, in putting financial values on the environmental aspect of their operations. In their 1991 accounts BSO/Origin attempted to give financial values to the environmental burden of the operating companies in their group and year on year are refining the process. Due to the specialist nature of their working practices, as an example it can only provide general guidance but it is a first in quantifying impacts.[18] No doubt this pioneering work will be followed by others.

Table 5.5 Development of the sustainable business report

Inclusion of statement in Annual Report and Accounts

Initial interest in the company's environmental performance is usually appeased by inclusion of an outline statement in the annual Report and Accounts; this is a low-risk and usually low-input strategy. The rush of consumer environmental concern in the UK during the late 1980s found a response by the inclusion of such statements, yet by 1992 a survey of the reports of 570 UK companies said that only 23 per cent mentioned environmental data and only 9 per cent disclosed concrete information such as policies.[15] A number of studies of the way in which companies impart environmental information have confirmed suspicions that many companies are using environmental disclosures as a channel for public relations and that the information is often partial, stressing only the best endeavours and not the underlying negative news.[16]

Environmental report

The creation of a stand-alone environmental report begins the process of identifying the environmental parameters of the organization. Often produced by drawing heavily on the environmental review prepared for the environmental policy development process, such reports tend to highlight positive data and focus on material that is easiest to collect. Yet these reports do identify a wider range of impacts and require more internal confidence in performance. For commentators, what is left out is often as useful as what is included.

Annual environmental reporting

To support the move towards annual environmental reports requires detailed and quantifiable data from the whole organization. In general this can only be achieved by a fully functioning environmental management system, thereby gaining access internally to all departments and ensuring consistency of information output. Initially, annual reports tend to be qualitative rather than quantitative; however, as annual reporting matures, detailed data such as Toxic Release Inventory information (TRI data) may be included, on a site-by-site basis, to support the overall conclusions of the report. (Some companies provide full TRI data on disk or on-line if required.) Reference to the annual environmental report might be made in the financial Report and Accounts, with greater detail provided on contingent environmental liabilities and a cost and savings breakdown for environmental protection and clean-up.

Stakeholder reporting

This type of report identifies the environment as one of the stakeholders of the organization; it creates an opportunity to identify the interests of other stakeholders in environmental issues as well as highlighting the social justice issues which are related to environmental damage in other countries, such as monoculture replacing biodiversity of species due to distorted trading patterns. Stand-alone Social Audits have some precedence in the USA in particular, although the European example of Traidcraft plc[17] is interesting, for despite initial criticism for placing too limited emphasis on environmental concerns it had matured to show the potential for stakeholder perspectives being addressed and reviewed on an annual basis. There is scope to develop this methodology to incorporate the environmental communication and disclosure requirements of any business in new and innovative ways.

Annual environmental and social sustainability assessment

This is the ultimate pinnacle of reporting, identifying the progress towards sustainable business by recording changes to the social, economic, environmental and human resource dimensions of the operating practice. Here the process of impact and resource accounting as we have identified it is developing into Full Cost accounting. Full Cost accounting seeks to formally identify environmental costs for the whole organization, involving new performance measurements for the move towards sustainability (e.g. remediation measurements for the amount of CO_2 produced by energy consumption) putting financial values against environmental impacts.

Recognition of corporate environmental performance by the financial community

Reporting environmental activity, as well as meeting the communication needs of stakeholders, has the potential to influence the organization's ability to raise funds. This can either be in a positive sense by increasing interest in the organization from those interested in environmentally responsible investment for ethical or growth reasons. Alternatively, the impact can be negative by default, as organizations who fail to keep pace with changing expectations are perceived to be either behind the times generally or hiding some unpleasant and therefore expensive contingent liabilities.

ETHICAL FINANCE

There has been much debate about the future of the ethical investment movement and whether the growth in assets under ethical guidelines will continue. The numbers themselves are just part of the equation; the very availability of funds which screen organizations on their records of environmental, social and political actions poses questions about overall corporate behaviour and influences the general trading climate. Table 5.6 lists the criteria that ethical funds and other like-minded organizations such as local government, pension funds and charities are using to determine their investment decisions.

Table 5.6 Screening criteria operated by UK ethical and environmental funds

Negative screening criteria operated by UK ethical and environmental funds

Poor environmental record
Poor working conditions
Operating under repressive regimes
Tobacco
Alcohol
Armaments
Gambling
Animal exploitation, experimentation and furs
Nuclear
Sexually explicit or violent media
Complaints upheld by the Advertising Standards Authority
Drugs
Political donations

Positive screening criteria operated by UK ethical and environmental funds

High standards of environmental awareness
Promotion of employee welfare
Equal employment opportunities
Useful contribution to community welfare
Provides environmentally beneficial goods and services
Provides socially beneficial goods and services
Has good customer relations record
Makes substantial charitable donations
Involvement in specific sectors (e.g. medical and health care, education)

Source: Rob Gray *et al*, *Accounting for the Environment*, © Paul Chapman Publishing Ltd., London, 1993, page 194. Reprinted with permission.

In addition to specialist funds there is some evidence of overall changes in attitudes within the financial community to environmental concerns.

RELATIONSHIP WITH BANKS AND OTHER CREDITORS

After a series of legal cases in which lenders were sued for the costs of cleaning up contaminated land taken as security on a loan made to a company, the US Environmental Protection Agency has produced rules governing lender liability under the Comprehensive Environmental Response, Compensation and Liability Act (CERCLA or Superfund) which restricts the liability of a lender when land has been used as security against borrowings for the clean-up costs of removing hazardous wastes from the facility. Given the wide range of items deemed to be hazardous under US jurisdiction, this redefinition has been helpful, but the loopholes are such that many lenders are retaining safeguards such as preliminary environmental audits for customers who are thought to have potential environmental difficulties with their assets. In Canada similar changes have taken place.

In the EU there is less agreement on the delivery mechanism for ensuring clean-up, although the EU Directive on Civil Liability for Damage Caused by Waste will have an impact on the financial industry, as contaminated land and contingent liabilities require review of the asset base of many organizations with land holdings. Registers of contaminated land exist in the USA, Canada, Germany, Denmark and the Netherlands, and the issue is still under review in the UK. The implications are immense, as potential costs of clean-up are staggering. In the UK alone current estimates of the total cost of cleaning up contaminated land are in the range of £10–30 billion.

Does knowledge about the history of a site impact on property values? According to the 1993 Hillier Parker Environmental Survey in the UK, over 50 per cent of property companies undertake an environmental investigation prior to purchase and 17 per cent would not be prepared to take purchases further after discovering that a site was contaminated. Such reticence may well increase as the scale of liabilities becomes more widely known. The UK banking industry now realizes the risks of property-owning. When Barclays Bank called in the receivers to the Mountleigh property group, taking possession of various properties as a result of loan defaults, they learned that major assets like the Merry Hill shopping centre at Dudley in the West Midlands were contaminated.[19] The Merry Hill centre was eventually sold at a reduced price, but only after several parties withdrew after learning about the contamination concerns. In this case there appears to have been a significant variance between the attitudes of UK and US companies to handling such a risk. This difference may be due to level of knowledge about the potential costs involved in clean-up exercises. Standards in the Netherlands are the strictest in the world, and these are used by some US corporations as best practice. The costs of remediation there average out at £1,900–£3,800 per square metre.[20] The latest consultation from the UK government suggests that contamination threats should be met by industry-wide insurance schemes, land should only be cleaned up to a level fit for the proposed use and contamination should be signalled by voluntary disclosure by vendors. The implications of this are very important to those holding land assets, and detailed advice on the latest position should be sought where necessary.

CREDIT RATINGS

The environmental performance of the operation can also impact upon the credit rating given by Moodies, the US financial rating agency, and others, as companies can be downgraded in the light of their anticipated liabilities and risk.

PENSION FUNDS

Trustees of company pension funds or those companies with institutional pension fund investors would do well to consider the potential impact of environmental issues to change fund managers' attitudes to risk and return, should concerted challenges be made by policy holders to raise the issue of ethical investment.

5.6 Sustainability and financial strategy

Companies wishing to move towards more sustainable operations will need to plan for those changes. This will require the skills of the financial director as much as those of the environmental director. A significant part of that transition will be the management of the capital base of the organization, to free up the capacity to make the required investments as company plans or external operating circumstances dictate. Indeed, there may well be occasions when the company's environmental aspirations are not achievable within its existing financial situation, and the ability to be independent is likely to be a key determinant in the transition to sustainable business.

Achieving sustainability may involve investment in new processes and the purchase of new technology and people to support them. It could even require withdrawal from existing sectors and finding alternative, more sustainable operations as investment vehicles to be the building blocks of a new-style organization. These are the ultimate decisions which any organization has to make and they require financial strength and the ability to transmute into another form, without ending up devoid of support from the financial community and the investor base.

The capacity to operate wider disclosure, in particular full cost accounting, where environmental costs are given financial values, will only be possible with the active intervention of the financial director. We would draw your attention to the wider-ranging nature of the changes involved in moving towards sustainability, as outlined in Chapter 2. Each of these changes will have a resource dimension.

5.7 Alternative company financial strategies

As we enter the twenty-first century, there are much bigger challenges ahead for financial managers than just questions over reporting and disclosure. So, for the company that has reviewed its environmental impacts what are the strategic options? (See Table 5.7.) The following generic strategies identify some of the current approaches, with the caveat that corporate financial strategy from an environmental perspective is in its earliest stages of development. Like many of the operational strategies found in this book, choice of approach is often determined by the overall operating culture of the organization and forms part of a progression of sophistication as corporate environmental strategy evolves.

Table 5.7 Different financial strategies and supporting environmental management approaches

Strategy	*Issues and techniques*
Compliance-driven	Compliance auditing BATNEEC BPEO
Resource use reduction	Waste, energy and pollution accounting Supplier auditing (emphasis on bought-in environmental impacts)
Risk reduction	Environmental impact assessment and environmental audit Development of EMS to ensure maximum knowledge base and control BS 7750/EMAS LCA/Ecobalance studies for communication purposes Defensive monitoring Superfund concerns; contaminated land review procedures TRI reporting
Progress towards sustainability	LCA/Ecobalance studies Annual environmental reporting, progressing to more advanced forms as system matures Stakeholder access and education Remediation planning Diversification away from unsustainable operations Benchmarking participation and supplier training Buying-up of tradable pollution licences (to improve working practices of competitors)

Compliance driven

This is the strategy adopted by many cost-driven organizations, where the concern about cost reduces the strategic vision to the shortest possible payback period. The tendency is to justify change only in terms of legislative requirements.

However, in sectors where the environmental effects are of the highest profile, due to the sensitivity of the product, such as chemicals and the energy sector, the costs of failure can be so high that payback concerns about environmental investment must take second place. Such organizations are often forced into full public disclosure through compliance with registers such as the Toxic Inventory Register.

Resource use reduction

This strategy allows environmental improvement to be afforded from extending the cost control already exercised, as part of a cost-conscious culture, to new environmentally defined territories. Here the treasury function is dominant within the financial strategy, and with it the mentality used so effectively to minimize waste of money is used to minimize use of physical resources. This approach does require the creation of new investment criteria to account for cost reductions over time. Saving money should not be the ultimate consideration; there are significant environmental improvements that produce only a break-even on benefits and these should be encouraged.

Risk reduction

This strategy works to reduce the contingent liabilities of the organization by acknowledging how changes will increase costs in the longer term and threaten market position. This risk adversity forces recognition of a longer time horizon for risk and therefore creates the positive climate for investment and research and development. The emphasis here is on disclosure of information to mollify concerned stakeholders.

Business opportunity

This strategy is based on gaining competitive advantage by responding to environmental issues as a business opportunity through innovation or diversification. Companies which have expertise, exclusive technology, products or services that meet environmental needs, may well be attractive purchases to the organization which sees the growth potential in this sector and is unable, or unwilling, to develop the equivalent internally. In other organizations where effort has been expended in working out solutions to environmental problems, the opportunity may be selling this technology on or developing subsidiaries to market the skill base independent of the core business. This type of diversification through innovation would build the profile of the company, as well as altering internal attitudes to environmental investment, when this is seen as a strategic asset-building activity.

Progress towards sustainability

The financial strategy for those seeking long-term survival is to mirror a corporate leadership strategy and extend this to financial decision making. It is characterized by:

- environment being fully integrated into the company's business strategy
- long-term investment
- attempts to close the loop in process cycles to ensure break-even in the Natural Resource Accounts
- transparency of information disclosure, not for purposes of insurance but as part of ethics
- relationship with new environmental funds as source of long-term capital for the company
- building of the environmental performance of the business as an 'intangible asset' which can ultimately result in improved valuations (such as premium over assets earned by shares in The Body Shop International) (See sustainability octagon.)
- remediation planning and the development of full sustainability accounts.

References and further reading

1. Rob Gray, Jan Bebbington and Diane Walters, *Accounting for the Environment*, Paul Chapman Publishing, London, 1993.
2. *ENDS Report* 189, October 1990, pages 14–18.
3. Adapted from a paper given by Prof. Rob Gray to Brunel Environmental Management Programme.
4. *From Ideas to Action: Business and Sustainable Development*, ICC, 1992, page 288.
5. Gray, op. cit., page 54.
6. Centre for Environment and Business in Scotland, *CEBIS Environmental Insurance Checklist*.
7. J. Leggett, 'Who will underwrite the Hurricane?' *New Scientist*, August 1993.
8. See David Pearce *et al*, *Blueprint 3: Measuring sustainable development*, Earthscan, London, 1994 for a fuller account of the debate concerning replacement of ecological capital.
9. *Business, Accountancy and the Environment: A Policy and Research Agenda*, Institute of Chartered Accountants, London, 1992, pages 1–2.

10. The Amended Enterprises Act of Norway 1989 requires a report to include information about physical emissions and contaminants and plans for improvement and clean-up.
11. Amendment to the Environmental Management Act 1992 requiring companies to publish detailed annual reports on their environmental impacts and improvement strategies, as quoted in the *ENDS Report* 222, July 1993.
12. Robert Lewis and Jonathan Isted, 'Beware of the Dangers of Disclosure', *Environment Strategy Europe*, Campden Publishing Ltd, London, 1993.
13. CBI Environment Forum Conference,
14. See Deloitte Touche Tohmatsu International, International Institute for Sustainable Development and Sustainability Ltd, *Coming Clean: Corporate Environmental Reporting*, Deloitte Touche Tohmatsu International, 1993 for more comprehensive coverage of the issues regarding external reporting.
15. Company Reporting quoted in *Environment Business*, 9/9/92, page 5.
16. *Financial Reporting 1991/92*, Institute of Chartered Accountants, London.
17. Traidcraft Social Audit Report 1992/93, Traidcraft plc, Gateshead, Tyne and Wear, UK.
18. Professor Richard Macve and Anthony Carey (eds), *Business, Accountancy and the Environment: A Policy and Research Agenda*, Institute of Chartered Accountants, London, 1992.
19. *Contaminated Land Developments*, Information for Industry, London, November 1992.
20. Dr Richard Haines, Ecotec, quoted in *Contaminated Land Developments*, November 1992, op. cit., November 1992.

PART THREE
Handbook of functional environmental strategies

Strategic choice at the level of the individual business functions is almost as important as at the corporate level. It is not unusual for companies' strategic aspirations to be derailed by contrasting and contradictory decisions at functional or business process levels. These decisions may be driven by very different considerations from those informing corporate choices, but which may appear equally, or more, compelling to the senior managers responsible for particular areas of operation.

For this reason, and to allow a more informed discussion of corporate choices at board level, it is essential to consider the impact of environmental pressures on each of these business areas separately. Indeed, it would be possible, though not desirable, to construct a corporate environmental strategy by coordinating and harmonizing departmental strategies. The converse seems to us to be quite unrealistic. This is why this Part provides what we regard as the indispensable means for turning corporate strategy into a set of well-grounded actions.

The six chapters in this Part look at the choice of strategies for the following major areas of the business:

Chapter 6: Marketing
Chapter 7: Research and development
Chapter 8: Operations
Chapter 9: Procurement and logistics
Chapter 10: Human resource development
Chapter 11: Corporate communications.

6
Market strategy and sustainability

1. Market strategy is an essential component of any move towards sustainable operations. Existing influences on the market will be modified and supplemented by environmental ones. We consider the relevance of the environmental challenge for the following influence factors:
   - Geographic spread
   - Protectionism
   - Changes within the technological base
   - Government intervention
   - Innovation
   - Shifts in consumer attitudes
   - Demographic change
   - Level of economic development

   and look at the additional influence of:
   - Environmental capacity
   - Environmental consciousness.
2. We believe that deep changes to markets are taking place as a result of environmental concerns. Evolving a market strategy to meet these is assisted by traditional and new profiling techniques such as:
   - SWOT analysis
   - Sustainability octagon
   - Market maturity analysis
   - Life-cycle analysis
   - Environmental option assessment.
3. Environmental product strategy will involve a reappraisal of the traditional alternatives of diversification, innovation and re-positioning.

food exporters may be forced to feed their own populations first.

The need to protect home markets from more minor environmentally damaging incursions should not be overlooked. Countries legitimately want to protect themselves, either from environmentally substandard or damaging technologies and processes. The debates surrounding Northern European packaging regulations as a restraint of trade are one example. Concern about the ability to maintain high environmental standards featured heavily in anti-EU campaigning in referenda by some of the potential entrants to the EU. It is likely that the next GATT Round will be forced to deal with numerous environmentally inspired issues, as countries with a positive commitment to sustainable development attempt to manage the integrity of their borders.

Changes within the technological base

Historically, developments which bring about new transportation systems or means of storage also have significant implications for market size. The potential impact of new food handling technologies, including the much reviled food irradiation, could affect the ability of low labour cost countries to process food nearer to the site of cultivation, substantially reducing the potential for adding value in the consuming nation and thereby disrupting carefully constructed markets.

More likely in response to environmental pressures are policy changes that will involve the substitution of existing modes of transport and the reduced custom that results from people travelling less, representing a significant potential shift in market dynamics. Road pricing, higher fuel charges or carbon taxes and vehicles bearing their true environmental cost through greater taxation are likely to reduce the number of journeys making products and services dependent on transport less attractive due to higher costs. Recent changes to the UK planning guidance notes, which severely restrict the siting of out-of-town developments in the light of unwelcome changes to urban town centres, offer a foretaste of what is to come.[2]

Government intervention

For many on the political right the environmental agenda is seen as a siren call to governments to intervene in the market, through command and control systems or fiscal regimes, and one that must be resisted. Yet those in global businesses addressing the implications of sustainability, such as the members of the BCSD, believe that some form of intervention is both likely and necessary. The aim of these business leaders is *not* to avoid regulation, but to generate discussion of what constitutes the right *mix* of regulation, economic instruments and self-regulation. There is evidence of lobbying to influence the policy agenda of many governments. The choice that is made in each market will change its structure considerably.[3]

The issues concerning the role of businesses in the post-Cold War world are complex; what seems to be emerging among opinion formers is better understanding that business must *lead* rather than follow governmental action. In their view, some limits imposed by legislation are considered reasonable, such as restrictions on movement of toxic wastes, while other policy dilemmas are considered best met by self-regulation, such as the means of introducing cleaner technology more quickly and to the level of best available practice. Others believe that change is unlikely to happen in the necessary time-frame without the necessary framework being created by government.

Innovation

Finding ways to apply technological innovation to manufacturing processes is just one aspect of the management of environmental change; technological developments obviously have enormous potential to make markets. There is the opportunity to create new products by providing alternative means of meeting existing needs – such as photo-voltaic cells in energy production – or in providing consumers with options that they had never considered before – such as contact lens with UV filters, offering protection against ozone layer induced rising UV levels. The developments made in consumer electronics, for example, have the potential to liberate many people from the drudgery of daily commuting through the avenue of 'telecommuting', with environmental benefits from reducing energy use and congestion and enhancing air quality.[4]

Shifts in consumer attitudes

The marketing discipline prides itself on reading and responding to shifts in consumer attitudes, and there are many examples of the environmental dimension influencing behaviours and attitudes.

Rapid changes in purchasing can result from recalculations of risk potential. This was the lesson learnt by the UK egg industry almost overnight, when consumers became aware of the common incidence of salmonella in many eggs and 'cook chill' foods which they regularly purchased. Similarly, the impact on the US apple-producing industry of the Alar scare in 1988 brought about an immediate change in its fortunes, as overnight millions of tonnes of products were shunned in response to fears about the carcinogenic potential of the Alar chemical used in apple cultivation. The Alar scare alone is estimated to have cost growers an estimated US$100 million in sales.[5]

These may be harbingers of more radical attitude changes. Dislike of 'chemical' agriculture is a small but growing factor influencing consumer choice in the West. It is based as much on individual health fears as on the environmental message that top soil is being irretrievably damaged (at least in the short term) and persistent man-made poisons which cannot be naturally dispersed are building up throughout the food chain. With the growth of environmental awareness these perceptions are beginning to percolate from fringe groups into the mainstream, with incalculable long-term effects on the market. Recent UK governmental advice on the preparation of carrots to avoid excess chemical residues appears to be filtering through to concern about farming methods, the environmental impact on growing areas as well as human health.

Demographic change

The number and characteristics of consumers themselves also change over time. World population growth will transform many markets, and demographic change is likely to be a key dynamic of consumer marketing during the early twenty-first century.

The well-known long-term changes in the age profile of the population of most OECD countries will have major impacts across a wide number of sectors as consumption changes and need for services, alters as they age. In the EU the population is ageing: from an anticipated 17.15 per cent of population over 60 years in 1980 rising according to one estimate to 24 per cent by 2010.[6]

Understanding markets based on the physical needs, as well as the psychological perspectives, of an older population will test the strategic skills of many marketers. With ageing populations will also come changes in the dynamic between their role as the owners of capital and much of industry, through pension-fund and stock holdings and the interests of the working populations of the future. It is likely that spending and consumption will be somewhat less of a concern for people as they age than conserving capital and ensuring a sustainable income base and that the environmental credentials of products will be more important to them.

Level of economic development

In addition, the economic progress of the nation states one trades with, and works within, also has fundamental impacts on market sizes and development. Overall levels of wealth within a market change over time, and the level of development or quality of economic infrastructure directly affects the potential for market growth. Countries where an increasing proportion of national wealth is being absorbed into environmental remediation work may have less available for consumer consumption by their citizens.

Environmental capacity

To these traditional aspects of long-term planning we would add another dimension – *environmental capacity* – that is, the ability of natural systems to accommodate to human-made changes. Environmental capacity varies considerably according to the strength of each individual eco-system.

In many parts of the world there is growing evidence that the very structure of our support systems is under attack. The latest reports on the health of the North Atlantic, for example, suggest that trawling is removing so large a percentage of fish in certain categories that continued fishing is under threat. 'Some 50,000 people were put out of work in Canada alone when the country shut down the cod industry to allow stocks to recover'.[7] This does not bode well for consumer marketers, who have spent years building up cosy images of the sea as an endless provider of cheap mass-marketed foodstuffs, when there is a likelihood that many of the fish products will not be available due to lack of raw materials. Is the end in sight for the cod fish-finger?

Environmental consciousness

Environmental awareness varies considerably throughout the world economies, and there are often marked differences in what Françoise Simon calls 'green IQ' within regions[8] affecting the globalization of marketing communication and product strategies.

It is likely that as consumers, and society at large, become more aware of both the mechanisms that cause environmental damage and the results of that damage, the overall context in which other factors operate will be fundamentally altered. After the peak in the late 1980s awareness of environmental issues is maturing, and there is a real growth of understanding of the integrated nature of the problems, if not as yet their solutions.

The EC Eurobarometer study, tracking attitudes to the environment, has found that in every EU country since 1987 there have been substantial changes to the group of citizens who see concern about 'protecting the environment and fighting pollution' as an immediate and urgent problem.[9] (See Table 6.1.)

Table 6.1 Eurobarometer study: the European public and the environment

Q. 'Many people are concerned about protecting the environment and fighting pollution. In your opinion is this...?'

	Belgium	**Denmark**	**France**	**Germany**	**Greece**	**Ireland**	**Italy**	**Luxem-bourg**	**Nether-lands**	**Portugal**	**Spain**	**UK**	**EC 12**
	%	%	%	%	%	%	%	%	%	%	%	%	%
An immediate urgent problem	85	87	80	88	97	70	91	83	84	73	82	82	85
More problem for the future	11	10	17	10	3	19	7	12	11	14	12	10	11
Not a problem	2	3	2	1	0	8	1	1	3	1	1	3	2
Increase since 1987	**12**	**5**	**21**	**4**	**15**	**7**	**6**	**–1**	**19**	**9**	**8**	**15**	**11**
Base:	1 036	1 000	1 005	1 065	1 000	1 001	1 046	496	1 002	1 000	1 000	1 319	13 082

Source: EUROBAROMETER: Europeans and the Environment in 1992, Spring 1992.

This attitudinal shift has continued to grow since the late 1980s, and despite the negative impression given by many commentators commitment to environmental action has been maintained throughout the European recession of the early 1990s. In 1992 the Eurobarometer study was used to attempt to establish the degree to which EU (then EC) citizens were ready to balance economic considerations against their desire to improve the environment. The findings were surprising. While 69 per cent across Europe believed that 'economic development must be ensured but the environment protected at the same time', of those who did choose between protecting the environment and economic development 22 per cent said that 'concerns about the environment should get higher priority than economic development'. This growing group of citizens varies from country to country in its demographic make-up, but it represents a remarkable consistency across the whole of the EU.[10]

6.2 Choosing your strategy: profiling opportunities

The overall growth, shape and nature of markets is a historical phenomenon which strategists and business leaders attempt to read and influence so as to profit from that knowledge. We are convinced that there is a deep change in market conditions currently taking place due to environmental forces. The individual company's task is to read the signals well enough to make a profitable and long-term intervention.

Integrating strategic responses to the environment into overall business planning requires use of traditional profiling techniques to pinpoint the likely changes to one's market.

Strengths, Weaknesses, Opportunities and Threats

The basic market development questions raised in a review of the business operating environment need to be put into the context of the business unit and the products and services it currently offers. There may be more up-to-date techniques, but the mental discipline imposed by the class analysis of Strengths, Weaknesses, Opportunities and Threats (SWOT) is still extremely useful. Table 6.2, based on a fictitious car producer, offers an illustration of some of the strategic principles discussed.

Table 6.2

Strengths	**Weaknesses**
Internal – Research and development skill base – Management commitment to improved environmental performance – Financial strength of the company	*Internal* – Lack of employee commitment to environmental issues – Control over distribution is limited and distributors' commitment to environmental issues poor
External – Growing body of scientific research to support case for technology shift	*External* – No consensus within industry of the appropriate route for development of new engine and fuel technologies – Lack of government commitment to developing that consensus
Opportunities	**Threats**
Internal – Develop cleaner technology – Use project as a means of developing innovative capacity – New technology will also have benefits in reducing environmental impact of production process, will hasten conversion to zero-emission production	*Internal* – Potential resistance of shareholders to loss of dividends due to cost of investment programme and loss of markets' confidence, making company target for takeover
External – Anticipated high consumer demand to meet more stringent emission control legislation – Collaboration and development of partnership with suppliers and other manufacturers on technology upgrade	*External* – Government introduction of road pricing as macro-solution to overall environmental impact of cars, road-building and parking – Competitor activity suggests their chosen technology may gain dominance due to timing of market entry

The sustainability octagon

Peter James has adapted Michael Porter's concept of the value chain to interesting effect, conceptualizing the relationship between economic, financial and environmental sustainability in what he calls 'the Sustainability Octagon'[11] (see Figure 6.1). Adding to Porter's objectives of

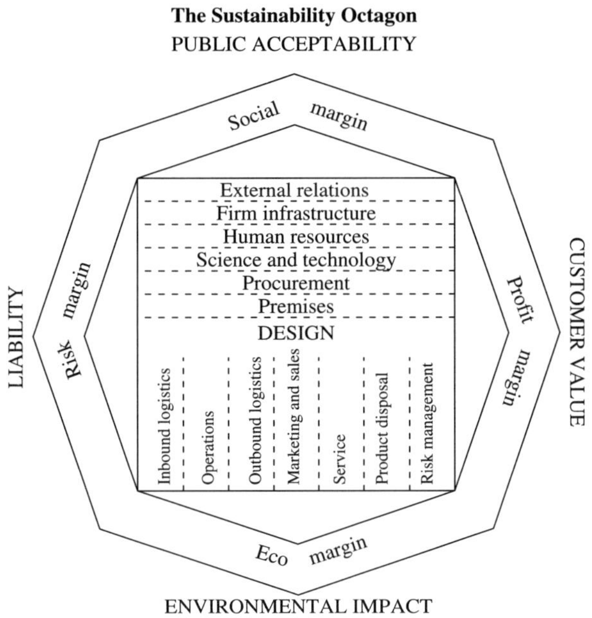

Figure 6.1 The sustainability octagon
Source: 'Greener Marketing: a Responsible Approach to Business', edited by Martin Charter. Reproduced by permission of Greenleaf Publishing.

maximizing profit margins, the James analysis explores other dimensions, including his new concepts of *risk margin, eco-margin* and *social margin.* Being able to articulate a value for environmental contribution to a brand's stature or corporate reputation allows for wider debate about the value of environmental activities to marketing strategy beyond consumer preference.

The *risk margin* is the ratio of actual or potential environmental liabilities to net assets. The *social margin* is the enhanced degree of trust or consumer franchise held by consumers in favour of the company contrasted against minimum operating requirements. The *eco-margin* is the ratio between a company's actual environmental impacts and the best possible level of environmental performance.

Although not precise measures, these ratios form a picture of the opportunities for differentiation over and above those traditionally articulated. In the absence of firm data some approximation is required based on the best available technologies and practices. Yet developing a wider analysis of the organization's actual impacts and its potential sustainability allows the worth of intangibles such as customer loyalty and commitment to non-profit-related values, e.g. being anti-animal testing, to have significance and create the space for exploration and strategic review.

Market maturity

In classic marketing, the Product Life-Cycle Analysis (as distinct from Environmental Life-Cycle Analysis) provides interesting information on which to base judgements about the performance of your product and its category. An understanding of the dynamics of the market situation is extremely useful in making judgements about the application of environmental concerns to marketing (see Table 6.3).

Table 6.3 Life-cycle of a consumer product

Stage one – Market development
This is when the new product is first brought to market, before there is a proven demand for it and often before it has been fully tested technically in all aspects. In sales terms demand is low and creeps along slowly.

Stage two –Take-off stage
At this point the market begins to grow rapidly as demand begins to increase.

Stage three – Market maturity
Demand levels off and grows much more slowly, at a replacement and new family formation rate.

Stage four – Market decline
There is less consumer interest as the item loses appeal, sales decline and market size contracts as customers move over to alternative offerings.

The objective in consumer-marketing is to develop products which will rapidly enter the take-off or growth phase, minimizing the risks associated with development. Market leaders often do not undergo the full product development cycle, leaving the innovation and risk-taking to others and seeking to capitalize on the pioneering work of others through imitation or outright purchase. To pursue sustainable development objectives there is a difficulty in utilizing such a strategy. The smaller company is unlikely to have the internal resources available to provide answers applicable to a larger competitor's quest for sustainability or, indeed, the ability to innovate by exploiting totally different technologies. When it comes to sustainability it pays to build from the bottom up.

Life-cycle analysis and sustainability

In addition to the traditional techniques outlined, there is a need to judge the risk associated both with your own product performance and the competitors' ability to reach sustainability for the longer-term health of your operation to be protected. The need to view any intervention in the context of market factors necessitates understanding whether there is a future for your type of offering as environmental factors change over time. For example, it may be that in the short term consumer demand will continue, but the legislative atmosphere in which the organization operates will curtail the exploitation of that latent demand. Alternatively, it may be that product benefit is dependent upon a type of material and characteristic that can only be generated unsustainably. In other instances, the product itself is predicated on post-use services being available, such as disposability and the assumption of free mixed waste disposal. Life-cycle analysis should include pre-production as well as post-use analysis of the product under consideration.

The car industry again offers an interesting example. While customers may want to continue to buy, and depend upon, private cars as a straightforward consumer choice, the need to reduce both the usage and number of cars, due to the environmental impacts of parking, acid rain and global-warming related emissions, ensures that this market is one which is highly likely to be constrained by outside forces. The UK's 1989 National Road Traffic Forecasts predict an increase in traffic (vehicle miles) of between 83 and 124 per cent by the year 2025 (on 1988); such increases in vehicle mileage would involve a total negation of the EU commitment to reduce carbon dioxide to 1990 levels, but would also represent a smothering in concrete of the countryside to a degree which most people would find unacceptable. Government intervention to reduce mileage is therefore highly probable, irrespective of individual consumers' preferences or party politics. It must be said that car manufacturers are responding with a massive R and D programme to find technology-based answers to the pollution impacts of the internal combustion engine through the use of alternative fuels and designs. Whether these efforts bear fruit, and mileage limitation can be fought off, is an open question. Constraints on the method of disposal of cars and their by-products like tyres are already with us in many markets, as governments are using a variety of means to ensure recycling and responsible end-use disposal.

Environmental life-cycle analysis is discussed in Chapter 7 and an example of its application is shown in the Scott Ltd example in Chapter 9.

Environmental Option Assessment (EOA)[12]

Environmental Option Assessment (EOA) is a new tool developed to aid decision making with respect to product selection, design and development, taking the whole product life cycle into consideration. It was originally developed to deal with a familiar scenario when working with environmental questions – attempting to interpret a maze of facts, many of them requiring specialist insight along with causal models too complex to understand easily.

EOA works from the premise that it is ordinary managers and stakeholder representatives, not off-line specialists, who need to make decisions. Any technique involving stakeholders, rather than professional technologists, must also be able to be understood and trusted if the decisions that result from the process are to be respected and therefore, accepted; EOA attempts to provide such a vehicle. Developed by the Association of the Dutch Chemical Industry, jointly with the Netherlands' Ministries of Environment and Economic Affairs and management consultants McKinsey & Company, EOA works from the standard Life-Cycle Management Model. This technique could be used to deal with areas of uncertainty when concrete decisions on new product specification and development and product substitution have to be made.

Development of a full understanding of the product under consideration follows this sequence:

1. Development of a volume flow diagram, clarifying the nature of the product as it passes through the six stages of its total life cycle.
2. Identification of the major environmental issues under different category headings.
3. Definition of the available options.
4. Prioritization of options.
5. Establishment of Action Plans.

STEP ONE – DRAW UP THE PRODUCT'S ENVIRONMENTAL LIFE-CYCLE DIAGRAM

Figure 6.2 shows the stages of the life cycle. At the outset it is necessary to chart the material flows into and out of each stage, including the inflow of raw materials, imports and exports, and especially the outflows due to emissions or waste at each stage. A particularly useful feature of the diagram are the four 'control valves' representing decisions about the flow of materials, products, emissions and recycling.

STEP TWO – IDENTIFYING MAJOR ENVIRONMENTAL ISSUES AROUND THE PRODUCT

The major environmental themes are first identified. (This can be done with the aid of a checklist.) With diverse groups and conflicting interests prioritization of the options available is crucial and here the innovation is seen: instead of a listing of personal priorities, each alternative policy or action is reviewed in the light of two aspects:

- Environmental profile
- Economic profile.

Environmental profile: What impact does each choice make on the

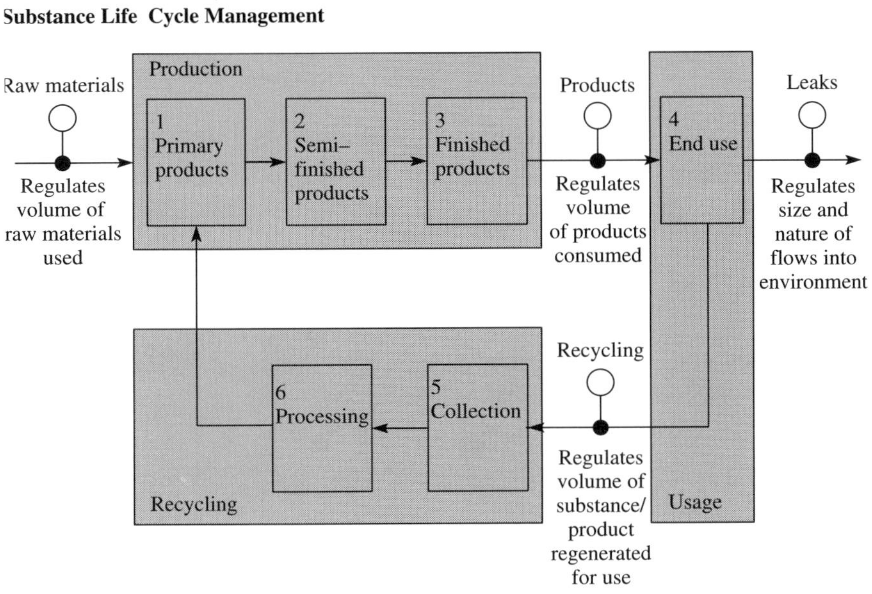

Figure 6.2 Step one: product environment life-cycle management

environmental themes already highlighted? The environmental profile represents the facts – hard and soft – that at the given moment are known about the environmental impact of an option.

Economic profile: This is the factual detail of the likely net economic impact of any option under consideration. Wherever possible it is recommended that economic effects are monetized and added up to indicate a total economic impact, including changes to such often overlooked areas as operating and capital costs.

STEP THREE – DEFINITION OF OPTIONS

This part of the process is to position the options, combining all quantitative values and qualitative views from the previous steps, to form a final viewpoint on the total environmental yield and economic impact. At this stage value judgements surface and most importantly if stakeholders are involved they themselves are questioned about their value judgements, either as individuals or as advocates for the organizations they represent. The strength of this methodology is that it provides an opportunity to create an 'option map' that is based on the fundamental concerns of all those with an interest and allows for a non-judgemental process for prioritizing the best ways forward.

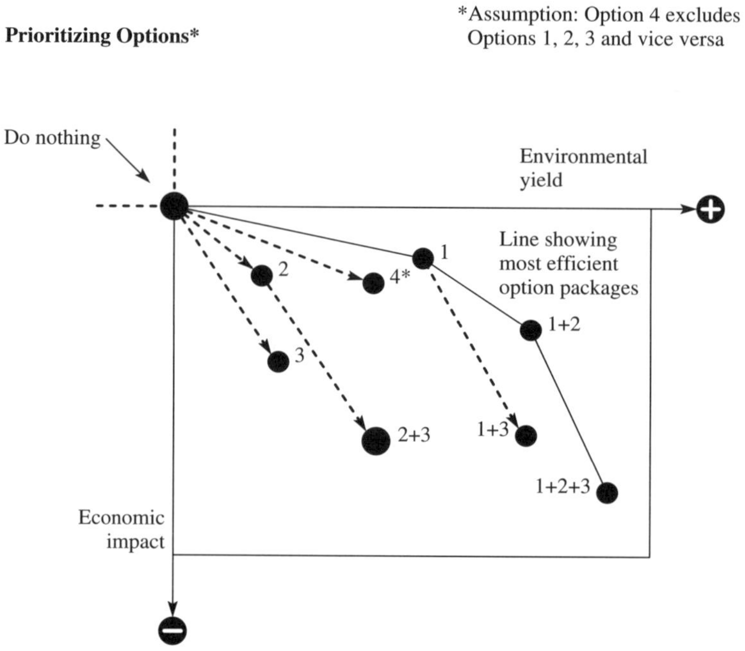

Figure 6.3 Option map showing prioritization of options

STEP FOUR – PRIORITIZATION OF OPTIONS

The options retained for consideration are plotted on a single 'option map' (see Figure 6.3) against an environmental and an economic dimension, which enable options to be prioritized and selected in the light of all the significant factors by a consensus of the decision-making group.

STEP FIVE – DRAW UP IMPLEMENTATION PLANS

Equipped with a list of prioritized options, strategists and managers are in a position to develop implementation plans to ensure that the strategic choice is realized. The ability to move from plan to action is dependent upon the level of commitment of the company. If the strategy involves third parties it is dependent on the resources input, the partnerships and degree of co-operation that are available and the level of behavioural change required.

The EOA process, like other models of consensus building, favours process over content in order of importance. Working with different stakeholders in any process is uncomfortable, and time and effort is required for people to settle into their new marriages of convenience. For companies with difficult stakeholder conflicts to resolve there is very little alternative to bearing this period of discomfort with good grace and learning to think about issues from differing perspectives.

EOA is a complement to market leadership. Its value is in allowing open discussion of often difficult and value-laden assumptions and reviewing the strategic options that follow from an environmental challenge. Related decision-making techniques can be applied to other environmental problems involving multiple stakeholders, diffuse benefits and impacts.

6.3 Meeting the product challenge

Understanding and meeting the issues raised by market analysis necessitates re-evaluation of both sector choice and product offering. In Chapter 4 we looked at alternative ways of approaching the strategic issues facing the business. In product terms these strategic choices can be characterized as one of the three following alternatives:

- Diversification
- Innovation
- Repositioning.

Diversification

Here it is either the choice of producing an offering which is complementary to your existing range, which can be produced from within your organization's own resources, or developing a parallel product to be created by buying in expertise through acquisition to create a new business centre.

The range of environment-related changes to consumer needs offers many opportunities. For example, in the small electrical goods sector there are gaps to be filled, offering solutions to concern about energy efficiency and environmental resources such as the quality of the air and water we consume. The Kenwood brand has responded to consumer concerns about the quality of publicly supplied drinking water with the development of electric water purification systems, as an extension from their traditional food-mixer ranges. In a similar vein, the UK high street retailer Boots has expanded its range of electronic products to market a range of battery rechargers and reusable batteries under its own brand name.

Innovation

Innovation involves the refocusing of available resources to fulfil an unmet need or the creation of a new technology which can be marketed. There is also the less costly route of imitation, applying solutions found in another sector to one's own or, as Theodore Levitt describes it, 'reverse R & D', where companies seek to create their own branded imitative equivalents of the innovative products created by others.[13] In the environmental area the concern is that one's competitors will not have completed the elementary work either, necessitating innovation. True innovation is a risky business, but one which does generate fresh concepts and ideas, meeting ordinary human needs in genuinely new and exciting ways – such as photo-voltaic roof tiles.

Repositioning

This involves re-evaluation to take into account changed environmental and consumer factors, so that by reformulation, representation and/or through the use of new channels of distribution the core values of your offering can be redefined to earn a new appraisal of your product by the consumer. (See Example 6.1.)

AEG Hausgeräte AG

AEG Hausgeräte is a major European manufacturer of white goods which has been developing an environmental dimension to its overall commitment to quality and design since the early 1970s. In 1976 the first washing machine with an energy-saving programme was introduced. In 1986 the ÖKO–LAVAMAT range of washing machines was introduced, designed to reduce environmental impacts by reducing use of detergents, water and energy consumption and increasing product longevity.[14]

Example 6.1 AEG Hausgeräte AG

The support given for energy efficiency information on electric consumer durables is another example of manufacturers utilizing improving product performance to reposition their reputations as trustworthy providers of consumer goods.

6.4 Marketing strategy based on environmental consciousness

Whichever product strategy is adopted it must be incorporated into a coherent marketing strategy (see Table 6.4). When approaching environmental matters there is a basic need for consistency between every part of

Table 6.4 Marketing strategy coherence checklist

Coherence checklist: A sustainable strategy

Product
- Does the product meet a real human need?
- Is the product as environmentally benign in use and disposal as possible?
- Can this be demonstrated through reference to scientific research validated by outside authorities and a comprehensive life-cycle analysis?
- Can you demonstrate best practice in production and general environmental management?
- Are the raw materials from sustainable sources which do not damage habitats or human welfare, and is the highest possible percentage of recycled material included?
- Has the product been tested by credible outside assessors to determine their views on the environmental claims?
- When the product was designed, was end-of-life recycling and repair potential maximized?

Price platform
- Does the pricing level selected reflect the environmental costs of bringing the product to market?
- Is the price margin consistent with others in its sector?
- If pricing differentials are required, have the reasons for this been explained to consumers?

Distribution strategy
- Have the logistics of the product been considered in terms of their environmental impact?
- Are alternative distribution channels available which might reduce environmental impacts?
- Does the chosen distribution channel provide adequate back-up to ensure that repair, recycling and reuse are possible?

Consumer proposition
- Does the consumer proposition present the total environmental performance of the product, ensure that there are no exaggerated claims being made?
- Does the proposition depend upon association with consuming more resources? (for example, speed in cars)

Brand or corporate values
- Do the brand or corporate values incorporate environmental responsibility as an extension of existing values such as 'caring' and 'safety'?

Communications strategy
- Are the media selected appropriate for the range of information required to support your environmental benefit and educate the consumer about the need for product benefit variation and or price differentials?
- Are messages consistent across media?

the overall marketing strategy. Failure to understand the environmental dimension may leave the organization open to criticism and ridicule. In the selection of marketing variables, unless there is consistency between operating practices, product benefits, brand values and the communication strategy adopted, the consumer is receiving conflicting messages. Even if the customer's own knowledge is insufficient to refute the presentation on offer, the lack of sincerity in an opportunistic, rather than coherent, position will be exposed, either by competitive interest or out of environmental concern by the pressure groups.

Examples of coherent and incoherent marketing strategies

The need for coherence in marketing strategy has been argued. But what happens when you do make environmental claims in business: how does the outside world view your position?

One of the few companies with a commitment to sustainable business is The Body Shop International, and we look at how it has responded at an operational level in Chapter 8. The Body Shop has worked hard to ensure that there is consistency between marketing activities and the values of the company. Yet even this advocate of environment-friendlier marketing has its critics. 'Dark green' environmentalists have queried the product offering itself, arguing that many of the 400-plus products are not actually required, and the packaging element of the company's gift baskets (a selection of toiletries in various types of secondary packaging) has come in for particular criticism as being wasteful. Given the importance of this item to the overall profitability of its stores, The Body Shop has tried to answer these concerns by introducing more recycled material into packing and ensuring that consumers receiving gifts are better educated about the reuse potential through literature included in the packs. Stringent efforts have been made to introduce good practices in internal processes to support the company's marketing.

Despite these mild criticisms The Body Shop International has been very successful; since its launch in 1976 it has grown to trade in 41 countries, employing 5850 people worldwide,[15] and much of this growth has been fuelled by the purchasing power of environmentally committed consumers. Above all, the brand is trusted and valued by the customer base as much for its consistency of core values as for the bright presentation of the product.

Scott Paper has also invested heavily in the promotion of its environmental performance for its household tissues and papers through support for 'sustainable' forestry and the view that tree planting reduces global warming potential. This profile has been the subject of attack by environmentalists such as the Women's Environmental Network who take issue with paper manufacturers over what they consider to be dubious claims for the 'environment-friendlier' nature of products such as kitchen tissue, which they consider unnecessary since reusable cloths and wipes would have less environmental impact. In addition, Scott Paper has been involved in several run-ins with the environmental movement over plans to develop large eucalyptus plantations in Indonesia and over its forest management policies in Nova Scotia,[16] and in both cases has changed policy in reponse to concerted campaigning. The company has received many awards for its creative approach to dealing with environmental problems and its products have consistently performed well.

BMW's efforts to offer environmental benefits to car purchasers in the form of recycling claims resulted in controversy. The claim of recyclability made in the early 1990s when facilities for decommissioning did not actually exist in many markets raised questions about the integrity of the environmental performance of the company and its willingness to provide improvements in environmental performance. This cast a shadow over the real efforts being made by the company, such as its plans to appoint regional dismantling and reprocessing centres, up to a planned strength of 10 by 1994. The dichotomy of a luxury car maker, with a product based on fuel-consuming speed and brilliantly engineered but polluting technology claiming environmental credentials, is an interesting one.

6.5 Pricing strategy and the environmental dimension

The issue of pricing and environmental performance is a vexed one for industry, with many consumers having differing perspectives on what constitutes acceptable price levels. Companies claim, with some justification, that investing in environmental performance leads to higher costs which need to be repaid, while consumers, on the other hand, do not understand why products based on claims for less inputs such as organic or recycled materials should suffer a price penalty. The attitude changes according to the level of technical sophistication of the product and its category, and is also influenced by the degree of information that accompanies pricing.

There is evidence that a reasonable percentage of consumers are prepared to pay a price premium for goods that they consider to have a genuine environmental benefit in performance, but there is a great deal of cynicism as to the nature of many claims that have been made. For the majority there is a preference to buy 'green' products *if* the price differential were not so high.[17]

For modern retailers the trend is to harness price concern and environmental concern by creating own-brand 'green' alternative ranges, offering high-performing environmentally-friendlier alternatives. The ability to provide green product alternatives is increasingly important to overall marketing performance in the grocery retail sector. Examples of this are the highly successful 'Green Care' range from Sainsburys and Loblaw's of Canada's 'Presidents Choice' range.

For manufacturers faced with this degree of own-brand competition, there is a fundamental question as to how they can absorb the conversion costs to better environmental performance when trading in mature markets, where price tolerance is inelastic.

The washing detergent market has been changed fundamentally by the changes in packaging and product application pioneered by the rival companies of Lever Brothers and Procter & Gamble. It appears that the application of the concept 'less is more' by product concentration, minimization of packaging, reusable outer containers and provision of washing liquid applicators to reduce product loss has enabled both companies to maintain margins while improving operational impacts such as the adoption of non-phosphate product formulations across more markets.

6.6 Distribution strategy and environmental impacts

Like all aspects of long-term business planning, channels of distribution for one's products and services must be reconsidered in the light of environmental consciousness. The advent of Life-Cycle Analysis and cradle-to-grave stewardship necessitates closer relationships with distributors in an effort to ensure that the environmental care aspirations of the lead company can be met on the ground. The European chemical industry has been forced under EU Duty of Care regulations to maintain much closer surveillance on the companies distributing and using their products once they have left the factory gates. The failure of users to properly mix and handle products in use, if this leads to environmental damage, reflects not only on the user, for example a farmer, but also on the originating chemical company.

Changing relationships with retailers is another example of a shift in balance between the parties as manufacturers are increasingly forced to change their operating procedures to gain distribution. UK retailers have been leading the way to ensure that measures such as the EU Directive on Packaging and Waste do not result in them having to cope with the spent packaging materials of their customers and suppliers, leaving the focus on packer fillers themselves.

Contact with consumers may need to be reviewed in the light of use-to-disposal commitments leading to adoption of direct marketing as a means of improving communication and to make return and reuse feasible. There are significant advantages in a manufacturer maintaining direct relationships with customers as margins decline due to retailer pressure and increased environmental costs. Retro-distribution schemes, such as the return of nickel cadmium batteries after use, will change relationships with dealer networks and consumers alike. The increase in product longevity required to reduce impacts may well force manufacturers to consider leasing operations.[18]

A more radical proposal for a fully sustainable 'intelligent product system' comes from the Environmental Protection Encouragement Association (EPEA) in Hamburg. This envisages durables and some other products being classed as 'products of service', which would not be sold, but *licensed* to the customer with ownership being retained by the manufacturer. The latter would be responsible for ultimate disposal and would be obliged to recycle or reuse all components which could not be broken down by natural processes, or pay for their indefinite storage.[19]

6.7 Promotional strategy

Evaluating one's promotional strategy from an environmentally conscious position requires evaluation of the following:

1. The nature of the message.
2. The environmental impact of the delivery mechanism.

Nature of the messages given

The pursuit of sustainability poses questions about the nature of consumer demand and the role of enterprise in satisfying consumer wants. The aim of marketing is to satisfy the consumer at a price that leaves enough profit. But in trying to develop a sustainable business the question of whether the product or service fulfils a real need has to be considered (like the criticism of The Body Shop that it is a luxury company producing things that are not actually needed, regardless of how sustainably produced). Consideration of consumers' relationship with your offering is advisable. Do people really need what is on offer to lead healthy and socially effective lives? Or does the fact that your organization is able to make its products in a sustainable fashion make any consideration of the usefulness of what is being made irrelevant? Does taking market share from less sustainable competitors lead to a net environmental benefit?

In the past many companies have been found guilty of a cardinal sin of environmental marketing, singling out one environmental product benefit without analysis of the overall environmental impact of the product itself. The promotion surrounding the launch of lead-free petrol in the UK is one of the most blatant examples of this, with companies claiming to be offering 'green' petrol and cars. The whole issue of credibility is covered in Chapter 11.

Environmental impact of the delivery mechanisms

In addition to the message, the medium itself is an issue that cannot be overlooked. There are significant environmental impacts associated with mass consumer marketing that need to be considered as part of a strategic marketing review. The most obvious areas are direct marketing and sales promotion, by virtue of the materials they consume by unsolicited communication. For many marketers demand creation is viewed as essential to continued existence, but leading-edge environmentally-driven companies such as The Body Shop would disagree, claiming that waste reduction must include promotional media.

6.8 Leadership and marketing strategies

With the environmental and sustainability arguments reviewed, each company has to reach a decision as to the appropriate marketing strategy for its stage of development and current level of environmental strategy development. The following represent a cross-section of strategic responses.

Development of parallel brands and product offerings

For companies with heavy investment in brand identities and corporate reputations, there is a clear rationale in expanding the values of the brands to include an environmental dimension, while still managing market penetration through maintaining the existing non-ecological offering. This creation of *parallel offerings* has been extensively used to maintain the financial strength of the company while educating consumers to accept changes to product performance and character.

The marketing of solvent-free paints by ICI and others is an example of softening consumer resistance to product changes by maintaining the original offer and increasing the percentage of product sold of the more environmentally-friendly formulations over time. This strategy is also being followed by the successful Swedish furniture retailer IKEA, presenting environmentally-friendlier design alongside traditionally manufactured product lines. The ability to develop parallel brands is often determined by the degree of investment made in consumer education and brand support, particularly given retailers' resistance to losing space to weak second cousins with inadequate promotional support.

Marketing of corporate commitments

For other companies where the corporate identity is stronger than the individual products on offer, it is important to build the profile of environmental activity within the whole company. There are numerous examples in the UK, but the stance taken by two of the largest privatized industries, British Gas and British Telecom, shows the different ways in which this strategic direction might be approached.

British Gas has been in the forefront of environmental communication in the UK and the promotion of information about the environment. In 1990 it commissioned a series of booklets from leading environmental authorities to comment on key issues of the day and promoted these through a national broadsheet advertising campaign. British Telecom, on the other hand, has focused its efforts on improving internal processes in advance of external reporting and has highlighted its potential power as one of the UK's largest purchasers of goods and services by supplier auditing. Both companies have increased their profiles through environmental reporting and see consumer perception of their environmental performance as a contributor to their overall success and social support.

Incorporation of environmental values into overall values

For other companies such as retailers where the corporate identity is the product, the brand itself is the cohesive force within the organization offering substance to product claims and communication. For retailers such as Sainsburys and Migros competing for business from the environmentally informed consumer, integration of the environmental policy of the company into all aspects of its activity is critical, even if product choice may be channelled into own-label or 'deep green' branded niches.

Leadership: Use of environmental concern as the cornerstone of competitive advantage and/or identity

For the company moving towards sustainability by making environmental concern integral to business strategy, the adoption of the highest quality environmental thinking offers a leadership profile and creates within the organization a new cohesive force. Companies undergoing turbulence in markets where change has been great have found that environmental excellence has the potential to refashion the devolved organization around new and potentially transforming values. Fast-moving sectors such as the computer industry and chemicals have seen a range of companies including Dow, Rhone-Poulenc, ICL and IBM using environmental achievements to reposition their companies, promoting new corporate values across flatter organizational structures, devolution of power and greater control by line management, with corporate commitment to overall environmental policy and targets. Here the environmental concern offers cohesion in the face of pressures to fragment the organization.

For other companies, environmental stewardship has been harnessed as a differentiation, bringing much desired attention to smaller companies and providing a social margin on which to capitalize (see 'The Sustainability Octagon', above). The Body Shop is a small-scale cosmetics company in the overall scheme of this mega-global industry, yet through its founder Anita Roddick's brilliant profile-raising activities concerning environmental and ethical issues, the brand is known and understood in many markets worldwide and this has been achieved without the use of expensive advertising support. It should be noted, though, that difficulties in the USA suggest that the company may well be forced to rely on more traditional marketing strategies in the light of severe competition. Norsk Hydro has similarly raised its profile in the UK to a much greater degree than its size in this market would normally warrant through leadership in environmental reporting.

References and further reading

1. James Goldsmith, *The Trap*, Macmillan, 1994.
2. *Guardian* Environment, 17 March 1994.
3. Stephen Schmidheiny (ed), *Changing Course: A Global Business Perspective on Development and the Environment*, MIT Press, 1992, pages 29–30.
4. Francis Kinsman, *The Telecommuters*, John Wiley & Sons, 1987.
5. Jacquelyn A. Ottman, *Green Marketing*, NTC Business Books, 1994.
6. Background Report on Framework for Action on Public Health, Commission of the European Communities, London, (COM (93) 559), 1994.
7. Lester Brown, *State of the World 1994*, Earthscan Publications, 1994.
8. Françoise Simon, 'Marketing Green Products in the Triad', *Columbia Journal of World Business*, Fall and Winter 1992.
9. Eurobarometer: *Europeans and the Environment*, The European Union.
10. Robert M. Worcester, *International Journal of Public Opinion Research*, MORI, July 1993.
11. P. James, The Corporate Response, *Greener Marketing*, Greenleaf Publishing, Sheffield, 1992.
12. Pieter Winsemius and Walter Hahn, 'Environmental Option Assessment', *Columbia Journal of World Business*, Winter and Fall 1992. For a full description of the methodology see the brochure *Integrated Substance Chain Management* available from the association of the Dutch Chemical Association (VNCI), Vlietweg 16, 2266 KA Leidschendam, The Netherlands.
13. Theodore Levitt, *The Marketing Imagination*, The Free Press, 1986.
14. *AEG: A Question of Housekeeping*, AEG (UK) Ltd, Slough.
15. *The Green Book*, Body Shop International, May 1992.
16. Adams, Carruthers, Hamil, *Changing Corporate Values*, New Consumer, Kogan Page, 1991, page 443.
17. Neilsen, '*Targeting the Green Shopper*', paper given to Brunel Environmental Management Seminar, 1993.
18. I. Blair, Green Products, *Greener Marketing*, op. cit.
19. Paul Hawken, *The Ecology of Commerce*, HarperCollins, 1993, pages 67–70.

7 Research and development strategy

1. R and D functions in organizations can play a number of environmental roles, ranging from risk-avoiding to responsive or proactive – or they can choose to remain inactive.
2. The most widespread strategy anticipates threats to the company's ability to make and sell its products, by mitigating the environmental impacts of materials, processes or the products themselves.
3. Some companies harness their R and D efforts to responding directly to perceived longer-term environmental needs, through design, new materials and process development initiatives.
4. In a smaller number of cases R and D know-how is geared to proactively exploiting the impressive market opportunities thrown up by the environment, by developing new, less damaging materials, processes and products.
5. In the name of scientific caution and detachment, some researchers adopt a wait-and-see policy. This helps to bolster the defensive strategy of companies determined not to forego short-term gains for long-term security. How damaging such a policy can be is shown by the ozone example.
6. The environmental agenda for R and D includes:
   - being a main source of science and technology information on environmental issues,
   - development of environmentally acceptable process technologies,
   - meeting long-term environmental needs through product design and development.

An ecologically focused research and development department offers the best prospects for improving the success rate of environmental business strategies. R&D provides the essential information for the future introduction of new, directly effective instruments of corporate environment policy. (W. Hopfenbeck, *The Green Management Revolution*.)

7.1 What role for R and D?

In parallel with the generic corporate environmental strategies described in Chapter 4, it is possible to distinguish a number of roles which R and D can play in mediating the organization's response to developing environmental pressures. These roles are probably influenced by the corporate environmental strategy, but may also help to determine it. To the extent that in science and technology-based companies the R and D function, or its equivalent, sets the pace and direction of innovation, the attitude of the scientists or technologists may well be one of the main determinants of the actual strategic stance.

In the following pages we describe four broad strategic R and D approaches towards the environment and the different roles it can play. The first three are not incompatible with each other but may be seen as steps in a progression of closer integration of R and D into the formulation of long-term business strategy.

Risk avoiding

On this assumption the basic role and starting point for R and D is to evaluate and foresee the risks to the existing business that are posed by the development of environmental pressures, particularly in the way of legislation, and guard against them.

Responsive

Beyond this risk-avoiding role, R and D can initiate a company response to the actual problems of the natural environment, over and above responding to legislation and competition resulting from other people's responses.

Proactive

Going one step further still, R and D specialists may be uniquely placed to identify new opportunities for the business from their understanding of the scientific and technical issues of environmental protection.

Inactive

Alternatively, as seems to be the case in some companies, R and D may take none of these roles, but hang back on the sidelines, expressing scepticism towards the environmental 'bandwagon'.

7.2 The risk-avoiding strategy

Among R and D departments that have taken on board the importance of the environment to their company, this is probably the most widespread approach, though it may be supplemented by one or both of the other 'active' roles. In this strategy, R and D takes on the role of environmental watchdog for the rest of the organization. It tries to anticipate threats to the organization's ability to make and sell its existing product range by keeping one jump ahead of legislation and customer perceptions of environmental danger. It maintains an ongoing review of products and processes based on the current state of the environmental argument.

Awareness and involvement

In order to play this role the R and D department has to maintain full awareness of current environmental thinking, particularly in relation to the effects of substances and technologies used by the company or which could be used in the future. In many cases this means taking part in the argument themselves, commissioning or doing research on impacts and alternatives, contributing to conferences and joint projects. In some cases companies set up special subdivisions of the R and D function to deal with environmental matters.[1]

Products

As well as monitoring the impact of processes and materials, R and D has to maintain a similar surveillance of issues concerning the company's products, both in use and after disposal. An ongoing awareness of legislation and public policy thinking, not only in the home country but worldwide, is equally important. Example 7.1 provides an instance of the extent and complexity of the surveillance processes for a transnational chemical producer. Anyone interested in the potential development of legislation should keep an eye on countries that have gone further along this road – e.g. Germany for waste control, the Netherlands for pollution standards, the United States for contaminated land.

Sensitive issues

There are also particular issues which have a high profile for certain industries and tend to occupy the attention of R and D specialists in that industry.

In the automotive industry, for instance, there are at present four principal focus areas for environment-related R and D: vehicle emissions, paint, fuel efficiency and vehicle recycling. These are at different stages of development – vehicle emissions are already subject to ever more stringent regulations in the USA and in Europe, which the makers have to meet to stay in business; solvent emissions from paint processes are at an intermediate stage of attracting increasing attention from the regulators; the questions of fuel efficiency and of ultimate vehicle disposal are beginning to surface as fields for possible legislative attention, but are still at a stage where research is seen as contributing to 'green' competitive advantage rather than to actually staying in business.[2]

Another major issue which has recently occupied a great deal of research time both in the chemical industry, as suppliers, and the electronics industry, as users, is the search for substitutes for CFC and other ozone-depleting chemicals used to clean printed circuit boards and for foam-blowing, in advance of the official phasing-out of production of these materials.

Life-cycle analysis

Another version of the risk-avoiding strategy is aimed at pre-empting the threat of lost market share due to consumers' concern about products' environmental impacts. This sometimes leads companies down the path of Product Life-Cycle Analysis (LCA) in order to enable them to claim that on the basis of the whole life cycle, 'cradle to grave' (or 'cradle-to-cradle' within a sustainable cyclical economy – see the schematic model in Chapter 6, Figure 6.2), a product is no more, or less, damaging than apparently greener alternatives. Life-Cycle Analysis is at present a rather complex and controversial technique, which can benefit from the technical sophistication of an R and D department, particularly when it is combined with exploration of less damaging alternatives to some of the life-cycle inputs, or is aimed at supporting an application for an ecolabel.[3]

Examples of LCA being used to attempt to defuse environmental criticism of certain products include disposable nappies (LCA commissioned by Procter & Gamble), detergents with or without phosphates (Albright & Wilson), milk cartons (the UK's Alliance for Beverage Cartons and the Environment). An EU study showed that company-sponsored LCAs tend to support the product of the company commissioning the study,[4] but they also provide a rational basis for public discussion and often lead to improvements being implemented by the sponsor.

Risk assessment

A more obvious aspect of a risk-avoiding strategy is, of course, risk assessment directed at accidents and incidents, and unintended contamination. Overall, these are more likely to be the province of operations management and the safety officer. However, some of the major disasters suffered by the global chemical, oil and nuclear industries do suggest a growing danger of unintended environmental damage through complex processes going wrong, and this is bound to become an increasingly important aspect of much of future process and product development.

Kodak Research

ASSESSING RISKS AND PLANNING ACTION IN A COMPLEX BUSINESS

Kodak's imaging business prides itself on its track record over many years of continuous improvement in reduction of environmental impact over a wide and diverse range of products sold both to consumers and to commerce.

The variables that Kodak has to consider in assessing environmental impact of their imaging products and processes and in deciding research priorities are both numerous and complex. Factors taken into consideration include:

- current and anticipated regulatory trends in all markets where the products are sold (almost every country in the world)
- competitive opportunities and pressures (reduced environmental impact as a product feature can be communicated to customers)
- environmental priorities of a widely diverse range of customers (e.g. different regions focus on different constituents of photo process effluent).

Cross-business and functional teams consider processes for prioritization of environmental goals as part of continuous product improvement. One type of analysis is based on regulatory forecasting – considering chemical categories (non-biodegradables, chelates, etc.) involved in the design and use of products one by one. For chemicals in each category, opportunities for environmental improvement and a variety of potential regulatory scenarios (banned, restricted, etc.) are considered and the latter are assigned probabilities, which can then be combined with the business implications, in the form of costs or revenues at stake, to derive a Pareto analysis of priorities.

A preferred analysis, based on Kodak's corporate strategy for minimizing environmental impact, expands an established hierarchy of approaches to guide research activities in relation to specific chemical categories, e.g.:

eliminate
reduce

recover, regenerate, reuse
recycle (recover and sell on)
treatment for safe disposal
environmental fate and effect studies (i.e. what is environmental effect if disposed of?)

Again, teamwork across businesses and with internal environmental experts helps decide priorities for research for each category of chemicals identified, taking account of regulation, competition, cost and revenue impacts. This is a valuable exercise as the type of research work required is different for, say, complete elimination and replacement of a chemical as against reduction, when maximizing the efficiency of use is the primary objective. The former may well require a greater understanding of the mechanism of which the chemical to be replaced is a part, while the latter may be an issue for process redesign.

Strategies based on such analyses are implemented through formal commercialization processes for the introduction of new products. A series of 'checkpoints' used in commercialization programmes are used to check against environmental goals to ensure that new products do not cause unknowing breaches of the environment strategy.

Example 7.1 Kodak Research

7.3 The responsive strategy

Detergents and post-use impact reduction

Henkel KGaA is, among other things, one of Europe's leading detergent manufacturers. From as far back as the 1960s Henkel has been aware of the concern over the effect of phosphates, one of the major constituents of detergents, in causing eutrophication (oxygen starvation) of surface waters through overnourishment of algae. In the company's own words:

> Due to the manifold tasks of phosphates in detergents on the one hand and the high quantities entrained into the environment on the other, the requirements on both technical application and environmental compatibility were very high. It thus took 20 years to find a substitute and to ensure by means of comprehensive tests that it will not have detrimental effects on man and the environment.[5]

As a result, Henkel introduced zeolites, which do not cause eutrophication, into its detergents and now receives royalties from the many other manufacturers who have had to adopt zeolites to meet the demand for phosphate-free detergents.[6]

This is an example of what we call a responsive strategy. It goes further than risk avoidance, by attempting to develop a strategic response to a potentially unsustainable situation in nature. At the same time as making an opportunity out of a problem, it protects the company's position within its existing business by finding a sustainable way of serving its market.

The motor car and the environmental challenge

In the process it may be necessary to redefine the 'business we are in'. This was illustrated by the board member responsible for environmental issues at Volkswagen, when he said:

> If in 20 years we will move around in electronically-controlled and magnetically-elevated cabins, then Volkswagen still wants to be the leading supplier of those cabins.

In line with this aim, Volkswagen, like some other car makers, has been pursuing R and D on electric hybrid drive vehicles which might become attractive if a range of possible political measures to deal with city centre pollution and congestion become realities. Over and above this, Volkswagen are researching electronic information systems to help reorganize parking and create combined private and public transport systems, with a view to preserving the viability of the motor car in some shape or form.[7]

It is in fact not surprising that the motor industry should be looking beyond meeting short- and medium-term constraints on vehicle emissions, energy consumption and waste. Producing as they do the largest, most expensive, most polluting and most desired consumer durable, which is widely regarded as quite unsustainable beyond the next few decades, they are bound to look actively at much more fundamental redevelopment of their product in the search for sustainable solutions to the real needs of the market from which they derive their income.

Designing for sustainability

The responsive strategy attempts to protect the company's markets by looking ahead to the requirements of a sustainable business and developing sustainable products that respond to customer needs. In many industries this role is shared by the development and design functions.

AT&T has set up an integrated 'Design for the Environment' approach, with scientists and engineers at its Engineering Research Centre working jointly with designers to prevent future environmental problems by looking at the whole life cycle of new products.[8] Philips has established a similar system based on an environmental design manual (see Example 7.2).

applied. Usually, as in the cases detailed above and below, the company extrapolates from its existing product base, in the light of the demands of sustainability, to identify new product and market opportunities.

Diesel engine manufacturers, the Perkins Group, clearly see major opportunities for expansion of their market in the imperatives of sustainable development. Tony Downes, Perkins' Director of Engineering Concepts, identifies one of the most important drivers of future design and development as 'the potentially explosive need for mobile power from the emergent nations, making unimaginable demands on oil resources and on the planet's ability to absorb emissions ... These are fundamental challenges but they are not insoluble. Technology will always have some kind of solution. What it takes is time, money and the will to succeed.' He sees Perkins' R and D capability at their Peterborough engineering complex, 'one of the world's most sophisticated centres for engines research', as the company's major asset for grasping the opportunities created by this problem: 'These new engines will be light-weight, have better power densities, lower noise, cleaner emissions ... they will have a capability to use renewable and alternative fuels such as vegetable oils and natural gas.'[13]

Rethinking the product range

When a company develops products or processes that are less damaging to the environment, the fact that it improves its business as a result does not, as some would claim, invalidate the genuineness of its achievement – very much the contrary! In a market economy long-term environmental improvement can only be sustained by commercial success. The interests of the environment are certainly not served if the more responsible firms go out of business while the polluters flourish.

What does worry people, however, is the strategy of companies who side by side with their 'green' products are actively promoting decidedly damaging alternatives, as if the polluting hand did not know what the green one was doing. It is therefore of particular interest when a company decides to use the research behind a successful 'green' product to set about transforming the whole of its range and production processes.

Esprit International, an international clothing company, carried out an internal environmental audit of its garment manufacturing, with researchers looking at all aspects of products and processes, including fibres and fabrics, dyeing and finishing processes, buttons and trims. As a result, it developed a special line of 14 garments called the 'Esprit Ecollection' which was launched in the spring of 1992 through stores all over the world. This collection incorporated a set of advanced design principles, including such requirements as:

- minimizing use of artificial fibres
- organically grown cotton bought direct from certified growers and wool bought direct from small sheep farmers, to encourage sustainable agriculture and farming
- minimizing harmful chemical treatments of yarns and fabrics
- a process of 'low impact dyeing' developed by Esprit resulting in 75 per cent less dyestuffs being released into effluent and less water and energy use
- heavy metals eliminated from dyes, development of natural dye techniques
- recycling of unwanted sweaters and factory scraps into wool fabrics for blazers and overcoats, instead of landfilling.

As a spin-off, the company gained an enormous amount of free publicity by its Ecollection becoming the most publicized fashion story of the year, with more than 800 articles being published in nine languages, as well as TV and radio coverage. The design principles and processes developed for Ecollection are now in the process of being transferred to the whole of the company's manufacturing range.[14]

Problems into opportunities

Environmental problems can be turned into opportunities. Companies facing pollution problems for which no ready-made, off-the-shelf remedy can be found, may have to invest resources in developing a solution for themselves but, having done so, they may find that they have acquired a potential product that can be sold to others facing similar problems. Another type of opportunity is that of turning costly 'waste' into a valuable resource. Pollution, like any form of dirt, is simply 'matter in the wrong place' and as in natural eco-systems one firm's waste can become another's raw material.

Elkem, a Norwegian-based materials company, decided to develop a new type of air-filtering technology for removing dust particles, in order to meet required air emission standards. Having developed the technology, the company found it so effective it decided to develop it commercially and is now successfully selling it worldwide.[15]

In a number of industries researchers have turned their attention to turning costs into profits – the costs of disposing of industrial by-products as waste into the profits from selling them on as raw material. At Du Pont a by-product of making nylon, called hexamethyleneimin, was found to have a market in the pharmaceuticals and coatings industries – so much so that when demand exceeded supply Du Pont began to make it for sale.

One of the most extended examples of a system of 'industrial ecology' fulfilling the principle *waste equals food* has emerged at Kalundborg in Denmark. This includes a coal-fired power plant selling waste heat to three other enterprises, waste calcium sulphate to a sheetrock plant to replace mined gypsum, and fly ash for concrete and road construction, an oil refinery selling surplus gas, after desulphurization, to the power station and the waste sulphur to a chemical plant, and waste heat to a fish farm whose fish sludge becomes fertilizer for local farms.[16]

SWOT analysis for R and D

A fully proactive strategy for R and D facing the problem of environmental protection and enhancement could, in addition to these avenues, include a systematic scientific/technological SWOT (Strengths, Weaknesses, Opportunities, Threats) analysis:

Strengths: in-house expertise, in-house technology, market knowledge, existing products, processes, by-products, etc. Which of all these are relevant and can be applied to environmental problems 'out there' that people (companies, institutions, governments, communities, consumers) will pay to have solved?
Weaknesses: what are the gaps in the equipment and knowledge base that detract from the company's capability to address environmental problems in the marketplace now and achieve sustainability in the future? How can these gaps be filled?
Opportunities: what special opportunities to contribute effectively and profitably to solving environmental problems do our special strengths and our market offer us? Can we use the environmental challenge, as we might any other challenge, to deliberately upgrade the innovative and problem-solving capability of our R and D?
Threats: how do our processes and products 'threaten' the natural environment and hence what is the threat of a non-negotiable demand for sustainable operation to the survival of our business? What risks do the gaps in our capability pose for us?

7.5 The inactive strategy

This strategy appeals to scientists who pride themselves on their scientific detachment and have decided that the environmental issue is being vastly exaggerated by the media, by emotional campaigners and by interested parties, including some other scientists.

A sceptical approach

They point out, rightly, that most statements about environmental risks are oversimplified in order to be understood by non-scientists, and they systematically align themselves with the sceptics in scientific arguments about environmental effects. They pick holes in the evidence for environmental concerns of other scientists, governments, environmentalists and ordinary people, in the same way as they have learned to do in discussing the research publications of scientists in their own field with whose views they disagree. As new evidence comes in, they treat it with the same suspicion as the earlier findings, pointing out its weaknesses and calling for suspension of judgement until yet more research has been done.

Precautionary principle: Two meanings

They will tend to reinterpret the internationally accepted ‘precautionary principle’ as meaning not that we should abstain from risking further environmental damage by continuing with a suspect activity even though its harmful effects are not 100 per cent proven, but rather that we should not forego income from the activity or spend money on preventive measures until the need is 100 per cent sure.

A case for caution, but ...

There is obviously a virtue in sounding a note of caution against uncritical acceptance of every new scare story. The public arena in which environmental decisions are made is full of conflicting interests, of special pleading and of people believing what they want to believe. Having accepted this part of the sceptics’ case, it is, however, worth pointing out that there are more people and institutions with an interest in suppressing or discounting unwelcome news than in exaggerating it, and that an excess of scientific caution may result in necessary precautionary action being pushed back beyond the critical point for averting major damage.

The ozone example

This has almost certainly been the case for ozone-depleting chemicals. The first warnings of the effect of CFCs on the stratospheric ozone layer were published by US scientists in 1974. The US government banned the use of CFC in aerosol cans in 1978. The British Antarctic Survey discovered the ‘hole’ in the ozone layer (in fact a 40 per cent decrease in ozone) in 1984, and the scientific paper about the discovery was published the following year.

The Montreal Protocol on first stabilizing and then reducing production of CFCs and other ozone depletants was signed in 1987.

A complete CFC phase-out by 2000 was agreed in 1990, and the target dates have since been brought further forward. But because of the quantities of ozone depleters already released and their extremely long life, their accumulation in the stratosphere is expected to continue through the twenty-first century, with consequent damage to humans, animals and plants. The ozone story is remarkable for showing the most rapid worldwide response to an environmental threat in history. But it was probably not rapid enough.[17]

Thus an inactive strategy which argues for more research and a wait-and-see policy, while current environmental concerns are left to a middle management HSE function, may have impeccable scientific credentials, and may in the short term save firms money, but we would question its overall appropriateness to the current situation on both environmental and business strategy grounds.

7.6 Environmental agenda for R and D

For any R and D unit that has decided or been asked to go beyond an 'inactive' role in its organization's environmental affairs, the agenda of items requiring attention is lengthening. The following outline is intended to provide a handy checklist of some of the areas to be considered – no doubt, individuals will be able to supplement it with other company-specific issues, new concerns and items we have omitted.

Information

As a repository of relevant scientific and technical information, the R and D function will need to scan the available information media for developments in:

Environmental sciences: including environmentally relevant developments in members' own disciplines, but also in other relevant areas, such as ecology;
Environmental technology: with developments in both 'clean-up' and 'clean' technology contributing to current views of what constitutes BATNEEC;
Materials: substances which are, or are on the way to being, prescribed or proscribed, requiring special treatment or disposal, substitutes for undesirable materials;
Legislation: affecting both processes and products, both in those countries where the company operates and its markets. (While R and D is unlikely to be the main source of this information for the company, it is vitally relevant to the development of new products and processes to have an understanding

not only of the existing legal position but also of the likely evolution of environmental legislation worldwide, so as not to be caught out by foreseeable developments (as some car makers were caught out by unleaded petrol and some refrigerator firms by the ban on CFCs).)

What is possible/technological forecasting: keeping abreast of future possibilities is part of the 'scanning' mission of a unit which aims to provide its organization with an early warning system. This may mean looking at some of the more unconventional and 'Futures' literature as well as mainstream publications.

Processes

Process development often overlaps with the R and D and Production Engineering functions. If so, the R and D role is likely to be concerned with longer-term developments and new departures in process technology. Issues of environmental impact are relevant to most aspects of process development, including:

Water: usage and contamination, including treatment and possible recycling of effluents.

Air emissions: prevention if possible, reduction and clean-up if not.

Waste, land contamination: elimination, reduction, recovery, recycling, disposal of toxics.

Testing: environmental impacts, avoidance of animal testing.

Cleaning: use of solvents producing volatile organic compound (VOC) emissions, elimination of ozone-depleting solvents.

Energy usage: CO_2 and cost saving, use of renewables, biomass, waste for energy.

Products

Product development may involve Marketing or a special New Product Development (NPD) function. As well as the perennial issues of appearance, quality, cost, functionality and fitness for purpose, a new set of environmental criteria has to be considered:

Materials: in many companies, R and D and/or the environmental manager have drawn up lists of prohibited and undesirable materials which are to be phased out. If such lists do not exist, one of the first tasks may be to create them, together with surveillance and review procedures (see the Kodak example above).

Packaging: a succession of consumer booms led to an orgy of overpackaging which has claimed increasing shares of raw materials and landfill space. The reaction from consumers and governments has pushed packaging to the centre of environmental attention and has led to R and D involvement in specifying and developing environment-friendlier and more economical solutions. The use of recycled material in packaging often forms part of these.

Impact in use: the most visible part of the product life cycle. Does the product pollute, create nuisance or noise, consume resources in limited supply, cause harm, damage or danger to people, animals, plants, ecosystems, the visible environment? Can any of these impacts be lessened by attention to design?

Energy consumption: an aspect of impact in use, receiving increasing attention from governments and customers in view of its global warming implications.

Durability, repairability: environmentally responsible manufacturers are beginning to move away from concepts like 'planned obsolescence' to build durability into their products and to make it easier to repair and replace worn parts.

Design for disassembly and recycling: total 'product stewardship' is seen as including the closing of the cycle at the end of the product's life, with 'cradle-to-grave' changing into 'from conception to resurrection'.

Ecolabelling: criteria for awarding ecolabels may include any or all of the above.

References and further reading

1. For example, Zeneca's (formerly ICI) Environmental Research Laboratory at Brixham.
2. Under US Corporate Average Fuel Efficiency (CAFE) Regulations, average fuel efficiency over a car maker's output has to meet a fixed minimum. Similar regulations are under consideration in Europe. Vehicle recycling is on the agenda in Germany.
3. At the time of writing, the criteria for granting ecolabels are based on LCA, and finding agreement within the EU Commission on appropriate criteria has been one of the reasons for delay in the implementation of the scheme.
4. *ENDS Report* 219, page 26.
5. J. O. Willums and U. Golüke, *From Ideas to Action*, ICC, 1992, page 292.
6. A recent life-cycle analysis commissioned by Albright & Wilson, leading manufacturers of phosphates for detergents, threw doubt on the environmental value of zeolites, but the findings have been strongly disputed (see *ENDS Report* 228, January 1994, pages 29–31) and the number of brands replacing phosphates with zeolites has continued to increase.
7. Willums and Golüke, op. cit., pages 120 and 150.
8. Willums and Golüke, page 106.
9. Both examples given by Willums and Golüke, op. cit., pages 196–197.
10. In the recent survey by Brunel University and the Prospect Centre mentioned in Chapter 4, 55 per cent of 80 strategic, environmental and general managers from a wide range of major companies endorsed the statement: 'Our approach

to environmental protection is based on the belief that unless we can make our business environmentally sustainable it will not survive.' (Fonda and Knight, 1993, op. cit.)

11. B. Good, *Industry and the Environment – A Strategic Overview*, CEST, 1991.
12. Study carried out by Ecotec for the UK Department of Trade and Industry, *ENDS Report* 230, March 1994.
13. T. Downes, 'Perkins put diesel in the driving seat, *Environment Strategy Europe*, Campden Publishing Ltd., 1993.
14. From Dan Imhoff, 'Esprit Ecollection: Ecology, Business and Community', *Environment Strategy Europe* 1993, Campden Publishing Ltd.
15. Willums and Golüke, op. cit.
16. P. Hawken, *The Ecology of Commerce*, HarperCollins, 1993.
17. See D. H. Meadows, D. L. Meadows and J. Randers, *Beyond the Limits*, Earthscan, 1992, for a blow-by-blow account of the ozone story.

8 Operations management strategy

1. 'Operations Strategy' is not the contradiction it sounds, for in many companies the Operations function drives both planning and design, and 'strategy' is a rationalization of reactive decision making. For the most polluting industries Operations are the key to environmental performance.
2. Operations Management is faced with a number of strategic issues.
3. The first and most basic issue is where to position the operational unit, on a continuum from reactive compliance to fully sustainable operation. Examples are given of the main strategic options available.
4. Environmental protection can be accepted as an unwelcome constraint, in conflict with efficiency goals, or it can be adopted as a positive stimulus to increase overall long-term effectiveness.
5. Many environmental improvements save money. Should the criterion for adopting them be how much they save (or how little they cost), or should the aim be to make the largest environmental improvements as cost-effectively as possible?
6. A further strategic decision is whether to seek independent validation of the environmental management system (e.g. through BS 7750 or EMAS). Linked to this is the question whether environmental performance information should be freely disclosed in the public domain.
7. Operations directors must also decide to what extent they are willing to collaborate openly with other organizations, inside and outside their industry, in a joint effort to speed up environmental improvement.
8. The basic choice in operations strategy is summed up as being between a defensive or a proactive response to the environmental challenge.

8.1 Operations strategy – a contradiction in terms?

Operational vs strategic

Strategy is often seen as the opposite of operations, like planning and execution, theory and practice. While operational management keeps things going in the present in accordance with existing missions, goals and strategies, strategic management is supposed to be busy thinking about the future and deciding in which direction to move the business. In a hierarchical perspective, strategy tends to be equated with senior or board level, operations with middle management. Thus, at first sight, to talk about strategy for operations management looks like being guilty of self-contradiction.

In practice, however, the strategy/operations model often seems to get turned on its head. In many organizations for much of the time practical people are getting on with the day-to-day work and responding to external changes as they arise. The business 'strategy' is a way of rationalizing what is already happening.

In such 'operations-driven' companies, a favourite environmental 'strategy' is to make environmental protection the responsibility of a specialist function (often coupled with safety and health) in order to protect operations management as far as possible against being distracted from their main task. Under these circumstances it is rather unlikely that the company will think strategically about joint optimization of operational and environmental needs for future business success.

Companies committed to process re-engineering or empowerment are even more dependent upon a clear and understood operations strategy, as the decision making within their organizations is more diffuse and all participants in a process have to understand the sustainability and environmental dimension.

Operations and sustainable development

If a business is genuinely planning to achieve sustainability it can hardly afford not to have an environmental strategy for operations management. It is one of the critical arenas where sustainability will be achieved or forfeited, the others being design and logistics – for what is at present unsustainable about our industrial systems is the way we make things (operations), what we make them of (design), how we transport them (logistics) and how we use them (based on design).

Operations and design – who drives whom?

Theoretically, the various design functions (R and D, Product and Process Development, Plant Design, Production Engineering, etc.) working in the context of corporate, financial and marketing strategies determine what shall be done, and Operations does it. However, the centre of gravity of the sustainable development process is likely to be determined by life cycle. Where the major environmental impact is from the production process (as, for example, in some parts of the chemical industry), operational management has to be the driver of sustainable business, for minimization of environmental impacts ultimately depends on operational practice rather than strategy at the corporate level. (See Example 8.1.)

The point can be illustrated by comparing the implications of environmental sustainability for cars and computers. The main environmental impact of a car is in its use, that of a computer during manufacture and end-of-use disposal. In the former case the key moves towards sustainability will be governed by design decisions which affect propulsion system, fuel usage and exhaust emissions (and even strategic moves to enhance public transport systems), while in the latter decisions about manufacturing processes (on such questions as energy, materials and solvents) will have the biggest impact (together with the potential for upgrading).

Moreover, in the typical organization where actual decisions about manufacturing technology, plant investment, process and product development are the responsibility of functions outside operations, they still have to satisfy the practical requirements of operational management. Even in consumer industries and services, where design will tend to play a dominant role in moving towards sustainability, we are not, by and large, dealing with organizations 'designed by geniuses to be run by idiots',[1] and the strategic approach of operations management to their role will still be crucial to success.

Strategic operations management in action

Harihar Polyfibres produces rayon-grade pulp for the Indian textile industry. A change in market conditions led the company to undertake a major productivity drive led by its senior executive president. By a combination of process improvement, reduced use of chemicals, waste reduction, increased energy efficiency, 'good housekeeping' and resource conservation, Harihar achieved reductions in energy consumption, chemical consumption and effluent load by 55 to 60 per cent over a period when production increased by 20 per cent. The payback period for an investment of US$69.5 million was two years, and only one year for the 'good housekeeping' improvements.[2]

Example 8.1 Harihar Polyfibres

Strategic issues for operations management

In the field of environmental protection, if operations management is to do more than comply with legislation and respond to *ad hoc* pressures, it requires a strategic approach. Such an approach can be defined in terms of a number of critical issues as shown in Table 8.1:

Table 8.1 Strategic issues for operations management

- Positioning: where does the company position itself between minimal compliance and sustainable development?
- Synergy: should operational managers be seeking synergy between environmental and operational performance?
- Cost–benefit: how do they approach the cost–benefit equation of environmental improvement?
- Validation: should the company seek external validation of its environmental management systems?
- Collaboration: should it collaborate for environmental improvement across a wide range of organizations?

8.2 Positioning

The first question is where to position the enterprise on a continuum ranging from minimal compliance with legislation to seeking to operate in a fully sustainable manner. In most manufacturing and some service areas, this is very much a question of operational strategy. Many companies are at the compliance end of the scale, seeing the environment as an unwelcome distraction from their business aims. Very few are at the 'sustainable' end, but there are a number of proactive positions in between, beyond simple compliance with the regulators' demands.

Table 8.2 Positioning: Some options

- Uncommitted
- Compliance
- Compliance plus
- Levelling up
- Continuous improvement
- Raising standards
- Comprehensive programme
- Sustainable business operations.

Compliance plus: while legislation remains the benchmark, the company takes its legal responsibilities very seriously, going out of its way to emphasize its commitment.

> Exxon Chemical Company will comply with applicable laws and regulations. Where the regulations are silent or ambiguous, we will meet the spirit of the appropriate laws and set responsible standards for ourselves.[3]

Levelling up: the company adopts the most demanding regulatory standards to which it is subject in any of the countries where it operates and applies them to all its sites.

> All new plants will be built to standards that will meet the regulations we can reasonably anticipate in the most environmentally-demanding country in which we operate that process. (Sir Denys Henderson, Chairman of ICI.[4])

Continuous improvement: the company sets demanding targets for 'continuous improvement' in source reductions, use of recycled materials, reductions in emissions and waste and elimination of toxic or potentially harmful materials.

> This is a widespread practice among companies aspiring to environmental leadership. Tough reduction targets have been announced by, among others, AT&T, Du Pont, ICI, SC Johnson, Monsanto, Rank Xerox, 3M.

Raising standards: the company makes a conscious decision to take a lead by raising environmental operating standards in its field, beyond legal or customary levels, where this is judged appropriate.

> The most advanced standards are used to set the minimum requirement for all Rank Xerox operations, regardless of the state of legislative development in individual countries ...
>
> At Rank Xerox we have made our commitment to manage everything that is within our control, not just to satisfy environmental standards but to help raise them. (Bernard Fourier, Managing Director.[5]) (See Example 8.2.)

Comprehensive programme: a company-wide environmental improvement programme that attempts to cover all aspects of operations. This can get very close to a 'plan for sustainable development', and is nowadays likely to incorporate an aspiration towards it without necessarily spelling out what the goal consists of.

> One of the most interesting examples is 3M, in its recent move from its long-standing 3P (Pollution Prevention Pays) programme to what it calls 3P Plus (see Example 8.3).

Sustainable business operations: the company commits itself to move towards an operational system which will be fully environmentally sustainable.

> Our mission is nothing less than to forge a new and more sustainable ethic for business. We want our efforts to set a precedent for others. (Anita Roddick, The Body Shop.) (See Example 8.4.)

The following are examples of three manufacturing companies who have taken a very conscious decision about their positioning.

Example 8.2: Rank Xerox's 'waste-free factory' programme

Rank Xerox Manufacturing Operations have set up a goal for their manufacturing sites which they describe as the 'Waste-free factory'. The details of this are contained in a matrix of nine action categories and five levels of achievement (which can also be seen as stages towards the goal). Individual plants are scored against the target by identifying their level of achievement for each category according to the descriptions in the matrix and applying differential weightings according to category.

The categories (with their weightings) are:

Environmental leadership (10%)
Strategic planning (10%)
Environmental communication and training (5%)
Solid waste management (15%)
Post-consumer material usage (15%)
Energy conservation (15%)
Hazardous and chemical waste management (10%)
Air emission management (10%)
Water discharge management (10%).

The precise descriptions of the levels of achievement differ according to category, but the following typical targets give an impressionistic indication of the whole:

Level 1 ('Growing')	– Present position identified and quantified.
Level 2	– Projects planned and prioritized.
Level 3 ('Maturing')	– Pollution prevention programmes: in place and working.
Level 4	– Programmes documented and inspected. 75% reductions from baseline positions.
Level 5 ('Interim waste-free factory objective')	– Environmental protection and pollution prevention integrated into management at all levels. 90 per cent reductions from baseline positions.

In two of the categories, Post-consumer material usage and energy conservation, a theoretical maximum/minimum is identified in the early stages, and targets are expressed in terms of this theoretical level. Thus energy usage is within 25 per cent of the theoretical minimum (under present conditions) at level 2 and within 10 per cent by the time it reaches level 5.

For the first three categories, achievement levels are described in qualitative terms. In the context of this book the category of strategic planning is of particular interest. The five levels of achievement are described as follows:

1. Environmental data are routinely used for current process and/or product improvement.
2. Strategic plans that include pollution prevention are reviewed and improved at least annually.
3. Strategic planning process for the waste-free factory is supported by the following tools: life-cycle analysis, environmental risk and/or hazard analysis.
4. Process and product improvement plans are in place throughout the facility that support the waste-free factory goal.
5. Waste-free factory commitment is integrated into short-term and long-term business plans.

Example 8.3: 3M: From 3P to 3P Plus

3M – the Minnesota Mining and Manufacturing Company – first introduced its initiative for cost-effective environmental improvement, 3P – Pollution Prevention Pays – in 1975, 'to protect the environment while producing economic benefits to the company'. The scheme involved enrolling employees at all levels in the search for improvements which would save money by preventing pollution – either at source by changing the processes or products by which it was generated or, where this was not possible, by recovery and recycling of waste materials.

Over a period of 16 years 3P resulted in more than halving pollution per unit of production through over 3500 successful projects, preventing more than 575 000 tons of pollution, with cumulative first-year savings exceeding US$550 million.

Side by side with 3P, however, 3M was setting up a wide range of other environmental initiatives, and in 1989 the company decided to bring all these projects together with 3P into an overarching environmental programme which was called 3P Plus. As well as the original 3P programme, which continued unabated, the new programme included systematic schemes for:

Air emission reduction
Energy management
Resource recovery

Productivity
Environmental management
Internal compliance auditing
Reviewing environmental marketing claims
Employee environmental recognition
Specific projects for phase-out of ozone-depleting chemicals, asbestos abatement, polychlorinated biphenyl (PCB) elimination, upgrading of underground tanks and pipes, improvement of hydrofluoric handling and storage.

3P Plus also included a set of quantitative pollution prevention goals:

By 1993:	Cut air emissions by 70 per cent
By 1995:	Reduce energy consumption by 20 per cent per unit of production. Cut generation of waste by 35 per cent
By 2000:	Reduce all releases by 90 per cent Cut generation of waste by 50 per cent
Beyond 2000:	Approach zero pollution and achieve sustainable development.[6]

Example 8.4: Dedicated to sustainability: The Body Shop

The Body Shop is of particular interest as one of the first international businesses that has officially and publicly dedicated itself to the goal of achieving sustainability in its operations, and is systematically setting about implementing it. Its Mission Statement (August 1994) commits it, among other aims to 'ensure that our business is ecologically sustainable, meeting the needs of the present without compromising the future'.

This aspiration is expressed in The Body Shop's corporate, marketing, communications, financial and organizational strategies, but its fullest expression comes in the firm's operational strategy. What then are the major ingredients of a sustainable development strategy for operations?

Resources: the aim is to use only renewable resources. The main areas where this is not happening at present are fossil-fuel-based electricity (see Energy) and the plastics used in packaging. As regards the latter, the company aims one day to replace petroleum-based plastics by biobased polymers from renewable sources (the main difficulty in the past has been finding a supplier who does not test on animals).

Energy: the company aims to replace its use of fossil-fuel electricity by renewable energy sources, and to this end has invested in a wind farm which is intended, eventually, to supply to the National Grid at least as much electricity as the company is using. This is supplemented by an energy-efficiency project which has already improved efficiency by nearly 30 per cent on the company's main site. The site also provides some of its own electricity from its own wind generator.

Transport: the company uses a contractor who is experimenting with low-emission fuels, and currently runs vehicles on a more expensive low-sulphur diesel and has one gas-powered vehicle. (The Body Shop emphasizes that its aim in promoting fuel efficiency is not primarily financial but to reduce emissions of NO_x, SO_x and $C0_2$ and use of non-renewable resources.) The contractor's vehicles are also used to display messages that form part of the company's ecological campaigns.

Waste: in common with other environmentally aware companies, The Body Shop is committed to a four-tier approach: 'first reduce; next reuse; then recycle; and finally . . . dispose of waste using the safest and most responsible means available'. The way this policy is pursued in practice is illustrated by The Body Shop's approach to the packaging of its products.

Reduce: by February 1994, 13 per cent of products available from the warehouse were without packaging.

Reuse: to enable containers to be reused, a refilling service is available at all shops. Forty per cent of products are refillable, excluding colour cosmetics and accessories.

Recycle: by February 1994, as well as the 13 per cent of unpackaged products, a further 73 per cent (excluding colour cosmetics) were packaged in the 'most readily recyclable' materials used by the firm, thus exceeding the 80 per cent target of unpackaged and most readily recyclable (excluding colour cosmetics) for 1994. All shops take back containers for recycling and return them to the factory where they are shredded and passed to a plastics recycling company. These reuse and recycling facilities were promoted to customers by a three-week campaign in 1992.

Disposal: as well as waste minimization projects, a comprehensive overhaul of internal procedures and contractor vetting was undertaken and independently audited by consultants, who concluded: 'The Body Shop takes Duty of Care very seriously.'

Pollution: the company has confirmed, by three environmental audits, that it produces no significant pollution of its environment other than by its use of energy. The Chemical Oxygen Demand (COD) of aqueous effluent from the Watersmead site was targeted for 25 per cent reduction by 1994 and this was achieved. Further investment in water treatment is in hand.

Product stewardship (including purchasing): The Body Shop's programme of product stewardship has four components:

Supplier accreditation: all significant suppliers are star-rated according to predetermined criteria.

Life-cycle asssessment: great importance is given to ecological impacts of obtaining raw materials and ingredients, including effects on biodiversity, endangered habitats, human and animal rights and non-renewable resources.

Risk assessment: this involves close consideration of materials used in greatest quantities and those where there may be issues of ecotoxicity or biodegradability.

Guidance for buyers: detailed guidance notes on commodities and processes are issued to all buyers. These include a blacklist of substances and processes to be avoided in all cases.

Audits and targets: The Body Shop supports the European Eco-Management and Audit Regulation, particularly for its requirement of full publication of an independently verified audit report. In its published report, a comprehensive, informative and well-produced document, it sets out specific quantified improvement targets with deadlines and reports on their achievement in the following year's report.

8.3 Conflict or synergy?

Is environmental protection regarded as a burdensome constraint, or as a driver to improve effectiveness? Is productivity sought in spite of environmental requirements, or do you set out to turn environmental problems and targets into a positive impetus and opportunity for broader improvements in quality, cost and productivity, as a way of seeking synergy between environmental and industrial performance?

This question is the Operations Director's version of what is perhaps the main challenge to the business community – whether to turn environmental protection into an asset or a liability. Within a business-as-usual scenario, environmental measures are additional costs which have to be justified by arguments about image, corporate citizenship, 'licence to operate'. But a contrasting approach sees the situation in a very different light.

The main environmental damage inflicted by industry is in the form of pollution and waste, and the most usual way of dealing with it is by end-of-pipe clean-up measures. So first you waste materials which have cost you money, then you spend more money on cleaning up the waste. This kind of reasoning lies behind 3M's original 3P initiative, behind Dow Chemicals' WRAP (Waste Reduction Always Pays), Chevron's SMART (Save Money And Reduce Toxics) and many similar programmes which aim to achieve improved financial results simultaneously with reducing environmental impact.[7]

3M's programme has been going since 1975, and is stated to have saved more than US$550 million in first year savings alone. One among many examples of how such savings are obtained in practice was the introduction of an innovative solvent-free technology which both eliminated VOC emissions and got rid of the need to use expensive drying ovens, thus reducing total manufacturing cost[8] – a good straightforward example of a strategy of turning environmental problems into cost reduction opportunities.

Dow quotes similar successes[9], but it would be over-optimistic to suppose that under present circumstances pollution and waste prevention will automatically pay for themselves and produce savings. A *Financial Times* review of environmental action by the chemical industry[10] shows that many companies are concerned about the costs involved and sceptical about payback. The environmental director of Sandoz is quoted as saying 'It would be naive to say that reducing pollution always pays' in the context of the industry's 'Responsible Care' programme[11]

A report on Dow Chemicals' experience of implementing Responsible Care in the same survey also sounds a note of caution:

> Nevertheless, the overall cost component of Responsible Care is significant. Dow's environmental, health and safety operating expenses were US$600m in 1992 including US$200m for environmental capital investments. It hopes the investment will pay off in the long run by reducing environmental clean-up costs.[12]

Taken at face value, and divorced from the broader context of the 'environmental revolution', statements such as these might be thought to limit the validity of a 'synergy' strategy, which aims to enhance operational effectiveness in the process of improving environmental performance.

But the context is all-important. Dow's environmental director makes the point, as it relates to the chemical industry:

> Failure to improve both performance and image could be very expensive. Planning approvals for investment will be refused, making sites uneconomic. Plants will be driven further from employees and markets, adding further to costs. The chemical industry could be legislated to the point of economic non-viability.[13]

Given this context, a synergy strategy is not primarily about saving money against the status quo by cleaning up operations – though there are plenty of opportunities for this and many documented examples. But the main point is to use the environmental challenge to rethink manufacturing and other operations from the ground up, so as to replace processes that at present have to be either unsustainable or uneconomic by ones that are both cost-effective and sustainable in the long term as a basis for a successful and competitive business.

8.4 Cost–benefit of environmental improvement

Having decided that there can be real synergy between environmental protection and operational efficiency, because waste is expensive, there is then a great temptation to concentrate attention entirely on the cash savings that are to be had from pollution prevention, waste reduction, energy conservation and avoiding prosecution. This absolves managers from any suspicion of idealism or do-gooding, and confirms their hard-headed commitment to bottom-line business performance.

The real issue, however, is much less simple. In fact, this is one of the important strategic questions which operations management has to answer. Should cost and savings be made the major criterion of environmental improvement, giving priority to the measures with the most affordable cost or the biggest economic pay-off? Or should such savings rather be regarded as a welcome way of financing environmental improvement, while giving priority to the biggest *environmental* pay-off?

Rank Xerox state clearly in their Environmental Health and Safety Policy that 'environmental health and safety concerns take priority over economic considerations'. A number of companies are prepared to incur additional costs in pursuit of environmental objectives,[14] but clearly there is a limit to the costs that can be incurred in this way. Rhone-Poulenc's director of environment would probably be widely echoed:

> If waste minimization is not cost-effective, we simply won't do it. We have closed plants in the US because there was no point in investing to bring them up to standard.[15]

The strategic issue that needs to be underlined, however, is not how much a company is prepared to pay for environmental protection, but whether it will actively seek to make the most significant environmental improvements in the most cost-effective way, as against focusing predominantly on the costs and savings of solving problems of existing technology, usually by end-of-pipe measures.

On the face of it, practical business logic would seem to come down on the side of giving priority to the economic perspective. 'If it doesn't pay, we won't do it.' The main difficulty about this view is that the definition of what 'pays' is bound up with the time-horizon applied to the payback calculation. As discussed in Chapter 5, the usual economic criteria applied to investment these days tend to involve very short payback periods, which can be quite inappropriate when it is a question of making one's business sustainable in the long term.

> 'This return', [on proactive environmental risk management] says Robert Bringer, 3M's Environmental Vice President, 'is most likely a long-range return, as are most investments in research and development. Management, therefore, must

have the long-range vision and patience to invest now for future returns. There must be a recognition that cleaner products and processes will also require a long-range commitment of resources, since the development of these new products and processes may take 5 to 15 years. That kind of commitment demands courage as well as foresight, given the short-term financial horizons.'[16]

Some of the most significant measures, such as those designed to mitigate major climate change, to save forests, to avoid the destruction of long-term ecosystems or the exhaustion of topsoils, are of this nature. Their short-term return is likely to be moderate in scope and uncertain in achievement, but in the longer term they may well make the difference between survival or destruction of the business, the economy in which it is embedded and the life-support systems on which both depend.

Pollution control or pollution prevention?

This underlines what is probably the most fundamental strategic environmental issue for manufacturing industry: whether to adopt a preventive approach by investing in innovative 'clean technology' or limit expenditure to installation of end-of-pipe solutions to the problems of existing processes.

The changeover from solvent- to water-borne paints in some industries illustrates the difference. At Volkswagen, for example, conventional paint-spraying methods required filtration processes to reduce the very high concentrations of organic solvents emitted to the atmosphere (up to 80 per cent of the paint used) – a typical end-of-pipe solution. After extensive testing of a 'clean technology' alternative, a new process using water-soluble paint, Volkswagen constructed paintshops using the new technology at a number of sites in Germany, and intends to do the same at plants in other countries.[17] A more extended example of a clean technology approach is Example 8.5, shown schematically in Figure 8.1.

On paper there seems to be a fairly wide-ranging consensus in favour of pollution prevention at source, through investment in clean technology. The US government has even legislated for it in the Pollution Prevention Act of 1990, which attempts to give legal force to an environmental protection strategy based on prevention rather than remediation, by putting in place a number of facilitating and support mechanisms and requiring companies to report publicly on annual 'source reductions' of toxic chemicals (i.e. reducing or eliminating the use of these chemicals at the front end of the process).

On the other hand, a number of research studies have shown that clean technology investment is at present the exception rather than the rule.[18] Cairncross[19] suggests that this is due to end-of-pipe technology being the easiest and surest solution in the short term, even when it is not the cheapest. Pollution control equipment tends to be standardized, tried and tested, and available off-the-shelf. Its cost can be pinpointed more easily than unscrambling the pollution prevention component of a new process or plant. It is more 'visible' both in PR terms and often literally. In some countries it qualifies for government financial aid. End-of-pipe solutions can also be more attractive to the pollution control supplier, because they are more widely applicable than one-off customized pollution prevention systems and are often coupled with valuable maintenance contracts.

In theory, of course, the figures ought to speak for themselves. In those cases where prevention is a better bet the financial case should prove the point. Unfortunately, the real situation is not so simple, for two reasons.

First, clean technology solutions are usually innovative, involving greater uncertainty and longer lead-times. In the absence of a clean technology strategy they may not be considered or even discovered in the first place.

Secondly, on a straightforward Return on Investment (ROI) calculation, which discounts long-term savings and emphasizes short-term expenditure, the financial case for end-of-pipe remediation measures may well be stronger, in spite of the fact that putting off investment in prevention is reducing the company's longer-term ability to become environmentally sustainable.

This illustrates one of the major characteristics of environmental investment decisions – they cannot always be validated on a straightforward ROI criterion applied by accountants. They are truly strategic: they go to the heart of the future survival and competitive strength of the business.

Cleaner technology at Dow Chemical

Dow's pollution prevention philosophy calls for:

- optimal use of materials and energy
- avoidance of residues
- reuse of waste products
- using technology that minimizes pollution.

In its chemical plant at Stade, Germany, Dow introduced a series of measures to implement these principles and establish a system approaching a closed loop operation. The elements of this include:

- conversion of unusable fluid residues into valuable solvents by thermochlorination and chlorolysis
- thermal oxidation of the remainder to extract hydrochloric acid for use in the production process (5000 tonnes p.a.)
- biological purification of waste water

- sludge from this process being reduced by chemolysis (thermo-chemical hydrolysis) into biodegradable substances fed back into the purification plant (it is hoped to reduce an annual sludge output of 12 000 tonnes by over 90 per cent)
- high temperature incineration of solid waste, used to make steam (100 000 tonnes p.a.)
- incineration of exhaust gases.

The results include:

- 80 per cent reduction of atmospheric emissions
- substantial savings in waste disposal, water usage and energy, and utilization of materials that had previously been lost. (The return which these savings represent on an expenditure of DM 120 million plus is not quantified.)

(Condensed from W. Hopfenbeck, *The Green Management Revolution*, 1993)

Example 8.5 Dow Chemical

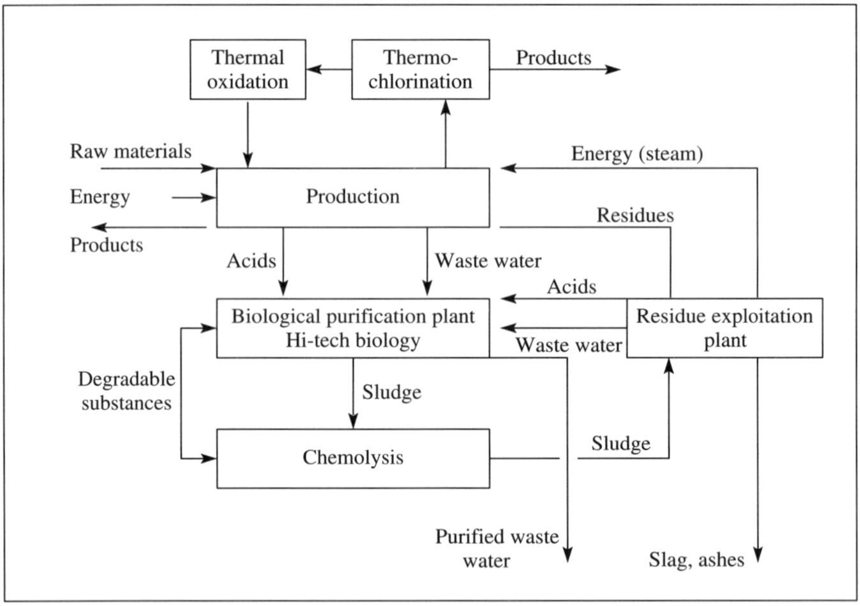

Figure 8.1: Dow Stade Compound concept production/environmental protection Reproduced from 'The Green Management Revolution' by Waldemar Hopfenbeck, 1993, by permission of the publisher, Prentice Hall International, Hemel Hempstead.

8.5 Independent validation

The decision to seek certification under the EMAS Regulation or the British Standard for Environmental Management Systems BS 7750 is a valid part of an outward-looking operational strategy, where it represents a public statement of environmental commitment and the willingness to build this commitment into management practices.

There are, of course, dissenting views. Some feel that BS 7750 is fatally flawed by its failure to set standards for environmental *performance* as well as systems, and by the absence of any obligation to publish environmental performance data. You can have the best paper systems in the world, it is argued, while you blithely go on polluting in practice.

EMAS, as well as being recognized throughout the EU, is seen to have the advantage (or disadvantage – depending on your point of view) of requiring the publication of an environmental statement verified by independent accredited environmental auditors. This makes it a more demanding commitment than the British Standard which requires independent accreditation only of the system, not the results, and does not call for publication. Like the British Standard, EMAS leaves the decision whether to set demanding or trivial targets to the individual organization. But by requiring publication of a statement of performance EMA relies, like the US 'right-to-know' legislation, on public opinion to discipline the laggards.

Some managers fear that getting involved with either of these schemes will expose them to a lot of bureaucratic form-filling and paperwork, as BS 5750, the Quality Management Systems Standard, is sometimes claimed to do. Others are worried about duplication or incompatibility with industry schemes, such as the chemical industry's 'Responsible Care'.

Many of these objections are well-founded. Nevertheless, the growing number of well-known companies who are introducing BS 7750 into their operations shows the overriding desirability of declaring environmental commitment in a concrete and verifiable way. Apart from anything else, such a move can confer an important competitive advantage when dealing with 'green' customers (and those who are themselves registered under one of the standards).

In any case, it is widely agreed that, if an organization is serious about its environmental commitment, then a systematic management approach is indispensable and it will have to include most, if not all, of the elements required by BS 7750 or EMAS, whether or not certification is sought. Without such a system, corporate environmental policies and strategies are liable to remain a dead letter, because there simply is no management process to translate them into practical reality or to monitor their effectiveness.

Publishing environmental information

The questions of whether to seek independent verification of environmental audits, and whether to publish the results as part of a policy of full disclosure of environmental impacts,[20] may be less straightforward. Some companies, such as Norsk Hydro and The Body Shop, have gone out of their way to obtain independent verification and to publish the results of their environmental audits in full. This has undoubtedly raised their profile as green companies and is likely to have furthered their business objectives. On the face of it, such public disclosure should be encouraged, and this is certainly the philosophy underlying EMAS. In the USA, it has been widely stated that the 'public right-to-know' laws, which oblige companies to publish full information on a wide range of emissions, have led those companies to become more active in eliminating them.[21]

Independent verification is closely linked to certification under EMAS or BS 7750, or to the desire to publish and be believed. But full publication is not necessarily every company's preferred option, and there may be good reasons for not doing so, other than having something to hide.

Under the pressure of legislation, public opinion and personal concern, many companies are beginning to take the environment seriously, perhaps for the first time in their history. To manage for effective improvement requires an unbiased management information system that gives the bad news as well as the good. But if environmental management has to be conducted in public, in the bright light of media attention, such systems can be dangerous and there is a risk of the PR tail wagging the management dog.

One scenario under these circumstances is that the main attention is focused, both internally and externally, on the aspects of performance which do credit to the firm and which, almost by definition, are in least need of improvement. Meanwhile the more dangerous practices are justified or explained away with all the brain-power and plausibility of hired experts until the company either believes, or becomes the prisoner of, its own propaganda.

The strategic manager who is concerned not merely with what makes good PR but also with the attempt to change the company's long-term chances in the sustainability stakes may therefore well hesitate before committing to a full programme of environmental disclosure. Public disclosure, if its genuineness can be assured, may provide a very effective sanction for pushing improvement through the company, but it may also lead to unpalatable truths being swept under the carpet. A common halfway house adopted by many companies who are either fighting off public criticism or aiming to establish environmental 'leadership', is to publish more or less extensive information about environmental targets and achievements as part of, or side by side with, their annual report. Well-known examples include ICI, Dow Chemical, British Airways, NatWest Bank and many more. (See Chapters 5 and 11.)

8.6 Collaborating for environmental improvement

Environmental protection is a relatively recent requirement for most businesses – we are all still learning what to do and how to do it. Sustainable business, as we have said, is not a fixed and known goal, but a new way of working which has to be invented. Collaboration between organizations helps to speed up the learning process for all concerned.

Benchmarking

As business has become more competitive, the practice of 'benchmarking' – comparing one's own performance with that of other companies on a range of significant indicators – has spread far and wide. Appropriately, it is now being applied to environmental as well as other kinds of performance. Next to the wishes of their superiors, managers are perhaps more open to influence from their peers, in their own and other companies, than from any other source. One of the most effective ways by which environmental management is spreading through industries is by peer and competitor pressure. By initiating, or taking part in, benchmarking exercises, the individual company both gains a better understanding of the standards of environmental performance to apply to its own planning and brings forward the day when environmental care will have become a normal part of good management everywhere.

Safety, Health and Environment management in the European chemical industry

In 1991/92 a questionnaire and interview survey of Safety, Health and Environment (SHE) management practices was jointly carried out by 18 chemical companies, covering six major topics. Conclusions from the survey, which were shared among participating companies, included the following:

Organization There were wide variations in the extent of integration of environment and health and safety management and in the resources devoted to them. Policies were developed at corporate levels, but performance standards were set on a regional basis, with only two companies applying the same standards worldwide.

Policies Almost all the companies had published policies, which included a commitment to Responsible Care and to working with the public on issues affecting them.

Training Perceived as a major priority and key to enhanced SHE performance. After auditing, the main area expected to require increased time allocation by SHE staff.

Performance measurement was more developed in the safety than the environmental area.

Expenditure and investment on SHE-related areas varied over a range, but were difficult to identify by many accounting systems. Environment-related expenditure was expected to increase.

Key issues and concerns for the next decade Product safety and regulatory changes, waste minimization and recycling, integrating SHE into line management, need for open information exchange and dialogue.

Example 8.6 Safety, Health and Environment management in the European chemical industry

An alternative to these outgoing inter-company exercises is to use published sources of information. Details of other companies' practice can be obtained from the growing body of literature on environmental management, from published company reports and from special surveys.[22] Obviously, these information sources are more 'edited' than the live information that can be obtained through the direct contacts possible in an inter-company benchmarking exercise.

Other forms of collaboration

Apart from benchmarking exercises and inter-company surveys of practice, there are many other possible ways of collaborating with outside organizations to achieve mutual benefits in the pursuit of environmental improvement.

Some of these are at the corporate policy-making level, such as participation in international and national joint initiatives like the BCSD, the ICC's Charter for Sustainable Development or the Advisory Council on Business and the Environment (ACBE), a group of chief executives set up to advise the UK government.

But many of the most fruitful collaborations are taking place at the operational level and are generating practical answers to urgent questions. When B & Q, the DIY chain, tried and failed to find a way of being certain that the timber they were buying had been sustainably produced, they participated in a collaborative project with a number of other companies who shared their objectives, and the Worldwide Fund for Nature (WWF). This resulted in the 1995 Group, a group of companies pledged to achieve criteria and a guarantee scheme for sustainable timber supplies by 1995 in conjunction with the recently formed Forest Stewardship Council. When some of the largest UK companies decided to introduce environmental purchasing policies they set up a joint working group within Business in the Environment (BiE – the environmental offshoot of Business in the Community) to work out the practical implications and produce a jointly endorsed scheme which would reduce the sheer diversity of environmental demands being placed on their suppliers.

There is a wealth of collaborative ventures for organizations to engage in, ranging from inter-firm collaborations like the two examples mentioned, and industry groupings like the chemical industry's 'Responsible Care' initiative or the motor industry's Automotive Consortium on Recycling and Disposal (ACORD), to collaboration with local authorities to set up recycling schemes, with local conservation groups to restore woodlands and wildlife areas or with educational institutions to sponsor and collaborate on environmental research projects.[23]

8.7 Strategic choice

The fundamental choice for operations management is between a defensive inward-looking approach which sees the environment as a threat to the efficiency and economy of the business, to be kept at bay by appropriate avoiding tactics, and a proactive outgoing stance, which pinpoints environmental performance as one of the key parameters of future success. It is at this fundamental level of operations strategy that the visions and aspirations of corporate environmental policies and mission statements are vindicated or buried.

The defensive strategy is most likely to be legislation-driven, though it may incorporate competitive considerations ('We don't want to be shown up or we might lose business'). The major emphasis will be on limiting emissions within permitted levels. Every survey that has been undertaken shows this to be the predominant outlook of today's managers.

The proactive strategy can be competition- and opportunity-driven and/or it can be an expression of a conscious sustainability strategy at corporate level. If the latter, the major emphasis will be on getting closer to cyclic, 'closed loop' operations and 'zero waste' targets. Either way, a characteristic of this approach will be forward-looking concern with the development of scientific knowledge, public attitudes and operational practice at the environmental leading edge, ensuring that the company is fully geared to build its future success in the new context of a sustainable economy.

References and further reading

1. Herman Wouk, *The Caine Mutiny*. The book opens with the statement 'The Navy was an organisation designed by geniuses to be run by idiots', a brilliant

summary of the old-fashioned 'strategic' design philosophy, which never had the remotest chance of working in practice.

2. S. Schmidheiny, *Changing Course*, 1992, pages 270–4.
3. Exxon Chemicals Operations Integrity Manual.
4. 'Investing in the Environment', speech given at *Financial Times* Conference 'Petrochemicals in Europe – The New Scenario', 28 November 1990.
5. Rank Xerox – Environmental Policy and Practices.
6. D. Benforado and R. P. Bringer, 'Pollution prevention and Total Quality Environmental Management', in R. V. Kolluru, *Environmental Strategies Handbook*, McGraw-Hill, 1994.
7. Bringer and Benforado, op. cit., list 24 such programmes.
8. Schmidheiny, op. cit., page 191.
9. See, for instance, the chemolysis process described in Waldemar Hopfenbeck, *The Green Management Revolution*, Prentice-Hall, 1993, pages 102–7.
10. *Financial Times* Survey, *Chemicals and the Environment*, 18 June 1993.
11. Ibid., page I.
12. Ibid., page VI.
13. Ibid., page I.
14. For example, S. C. Johnson Wax who were willing to pay higher prices for recycled material than its virgin equivalent at a time of supply difficulty, in order to maintain a programme of increased usage of recycled materials.
15. *Financial Times*, op. cit., page II.
16. Bringer and Benforado, op. cit.
17. J. Willums, and U. Golüke, *From Ideas to Action: Business and Sustainable Development*, 1992, page 168.
18. *The Promotion and Diffusion of Clean Technologies*, OECD, 1987; Irwin and Hooper, Clean Technology, successful innovation and the Greening of Industry: A Case Study Analysis, Business Strategy & the Environment **1**(2), 1992 ; Ecotec Research & Consulting Ltd, The Development of Cleaner Technologies: A Strategic Overview, Business Strategy & the Environment **1**(2), 1992.
19. Frances Cairncross, *Costing the Earth*, Business Books, 1991, pages 190–91.
20. A lot of this information should already be in the public domain, by virtue of the Environmental Protection Act (EPA)'s requirement for certain types of (regulated) emissions to be recorded in registers held by the local authority and accessible to the public.
21. For example, Rao V. Kolluru, *Environmental Strategies Handbook*: *A Guide to Effective Policies and Practices*, McGraw-Hill, 1994, page 19.
22. In the run-up to UNCED (The Rio World Summit Conference), the United Nations Centre on Transnational Corporations (later renamed Transnational Corporations and Management Division (TCMD)) ran a large-scale Benchmark Survey on Corporate Environmental Management which is summarized at some length in Kolluru, op. cit. A recent survey on business and environmental strategy was carried out by the Environmental Management Programme at Brunel University.
23. All three of these are among community initiatives undertaken by S.C. Johnson Wax in the UK. Like some other companies, Johnson has a budget earmarked for community projects, and has decided to use part of this for environmental initiatives.

9 Procurement and logistics strategies

Procurement

1. Large buyers in both the public and private sectors are increasingly using their purchasing policy to contribute to environmental improvement.
2. Where products that meet purchasers' environmental criteria are more expensive, the extra cost may be recouped by savings or absorbed by supplier, buyer or end-user. In the public sector 'value for money' may be reinterpreted in the light of published environmental goals.
3. Environmental performance is increasingly specified as a part of total quality. By working with their suppliers to help them achieve the desired standards, buyers can strengthen long-term relationships to mutual advantage.
4. To be effective, purchasing policy needs to be integrated with the environmental strategies of other functions.

Logistics

1. Contrary to widespread belief, a lot can be done by individual companies to reduce the environmental impacts of personal and freight transport.
2. One of the major impacts is emission of the greenhouse gas CO_2. This can be reduced by a long list of measures to increase fuel efficiency, reduce distance travelled and transfer to less harmful modes of transport.
3. The same measures will help to reduce another major impact: pollution from exhaust gases.
4. Refrigerated transport and storage still contribute to increased ozone-layer depletion – the replacement of these technologies is now a matter of great urgency.
5. Significant environmental improvements – and cost savings – can

be made by better waste management in secondary packaging, waste generated in vehicle operation and final disposal.

6. An environmental strategy for logistics has the primary objective of improving environmental performance, but can also produce cost savings. The main components of such a strategy are reduction and redeployment of transportation needs, fuel efficiency, replacing ozone depletants and reducing waste.

Purchasing, Procurement, Materials Management and Logistics are related and overlapping areas of management,[1] which link the organization to its commercial environment, interlocking with the other management functions we have been considering. The strategies governing them are as important for the company's environmental performance as are those for Operations, Marketing or R and D.

In addition, because procurement involves *relationships* with other organizations it focuses attention on the ways in which the effects of one organization's environmental strategies depend and impact on those of another.

9.1 Procurement strategy and the environment

In maintaining the necessary inward flow of materials and services to support a company's ongoing operations, the aims of a traditional procurement strategy would be likely to include the following:[2]

- minimizing overall costs (or maintaining them at a competitive level)
- securing required quality, specification and delivery
- developing effective and reliable sources of supply
- integration with other departments.

To this, recent developments have added two further aims:

- developing partnerships with suppliers to compete effectively in what is fast becoming 'a new ball game'[3]
- enabling the company to move towards ecological sustainability.

While an environmental strategy for procurement primarily addresses the last of these aims, it has become an important *dimension* of each of the others, as the following sections show.

In general terms, the purchasing policies of large buyers can exert critical influence over the actions of producers, often with environmental consequences. Thus one of the first questions for procurement managers to be aware of is whether the requirements they are placing on their suppliers are compatible with environmental protection. (For instance, it has been suggested that the demands of the modern supermarket for cheapness and regularity of shape, size and appearance of natural products have put pressure on growers to use methods that are environmentally damaging in the long term.)

The possibility, and desirability, of using their purchasing power as a means of influencing industry to take environmental protection more seriously, in the same way as consumers have been doing, is one of the major considerations that have pushed government, local authorities and an increasing number of large corporations to establish environmental purchasing policies. For instance, British Telecom, with an annual procurement budget in excess of £4 billion, decided that its purchasing policy provided the most effective lever for environmental improvement at its disposal (see Example 9.1).

BT's environmental purchasing policy

As the largest private sector purchaser in the UK with a total spend exceeding 1 per cent of GNP, with purchases in many environmentally sensitive sectors, BT has developed a comprehensive environmental purchasing policy which has strong top-level support. The policy is based on a generic environmental impact analysis across the whole product life cycle which is applicable to all significant contracts.

The company works closely with its top suppliers through extensive discussions and mutual presentations of environmental policies, concerns and improvement plans. Tenderers are given guidance and are asked to provide information about:

- manufacture (including subcontractors):
  - processes
  - materials
  - waste management
  - environmental hazards
- use:
  - hazards
  - power consumption
- recycling
- information and labelling
- suppliers' improvement programmes.

The 'return loops' of disassembly, reuse, recycling and disposal are fully explored during initial discussions, and potential future requirements for recycling and safe disposal are anticipated at an early stage, thus making the policy highly cost-effective, as well as enhancing BT's image and competitive position.

Example 9.1 British Telecom

To be effective, this kind of strategy requires close collaboration between the purchasing department and the company's environmental specialists, or even the injection of special environmental expertise into the buying function. For a green purchasing policy does not simply involve following a set of undisputed guidelines; it involves the making of informed *judgements* about controversial areas, it involves dialogue with suppliers, using their special expertise about the environmental angles of their own products and processes, appraising the quality of that expertise, and the genuineness of their motivation, where necessary by independent verification.

Financial and environmental costs

HOW MUCH?

How much extra should a company be prepared to pay for bought-in products and materials that meet environmental criteria? It sounds a simple question which could be answered in percentage or absolute terms. In reality the issue is much more complex than that, with answers ranging from zero to amounts that would make a given market non-viable.

An environmental purchasing strategy, like the business strategy of which it is a part, is ultimately directed at ensuring that the environment (and with it our survival chances) ceases to be treated as a free, expendable resource. It is a way of ensuring that someone pays the 'remediation costs' we have talked about in Chapter 2. The question is who? This is an issue which the strategy must address.

WHO PAYS?

The starting point for most businesses has been to ignore the question. They were not going to pay and they would not ask their customers to pay either. Someone out there would pick up the tab – the inhabitants of the regions from which the raw materials had been extracted, or people living near the factory who had to breathe the polluted air, or the local authority who had to clean it up, or our grandchildren who would inherit the impoverished environment we had created. Frankly, we didn't want to know.

Once this answer is seen to be unacceptable – and that is what environmental strategy is all about – the issue has to be faced. And it seems that there are at least five answers, which can be chosen either singly or in various combinations:

1. No-one need pay – because clean-up costs can be covered, or exceeded, by the savings from recovering and reusing pollutants and waste.

2. The supplier pays – by offsetting increased environmental cost against profit margins.
3. The buyers pay – out of *their* profit margin.
4. The customer/consumer pays – by increasing product price.
5. The community pays – by way of a government subsidy.

SELF-FINANCING

Self-financing the clean-up by recovering the waste should be everyone's favourite option, but it does require certain resources – at a minimum a person (or group) with the know-how and ingenuity to work out ways and means. In the context of a purchasing policy it is, of course, the supplier who needs to be encouraged (and perhaps helped) to achieve specified environmental quality by this means.

Buyers can refer their suppliers to well-known examples, such as 3M's 'Pollution Prevention Pays' campaign, which was reported to have saved the company over US$500 million over 15 years, counting only the first year savings on each project, as described in Chapter 8. Unfortunately, in the absence, so far, of well-established environmental accounting systems, it is difficult to judge the scope for self-financed cleaning-up across industry at large.

SUPPLIER PAYS

Even where self-financing is not an option most buyers will hope that their suppliers will be willing to absorb the new environmental costs in order to retain their business. Whether they do will depend on a number of factors, including their existing margins and marginal costs, the state of competition and how much they really want the business.

There is scope for negotiation, collaboration and creative compromise if both buyer and supplier take the view that they are dealing not with a one-off cost problem but with a necessary process of adjustment to a new industrial reality, which aims to make both parties more competitive in the new market context.

BUYER OR CUSTOMER PAYS

Large and powerful buyers may find that they have already squeezed their suppliers' margins to such an extent that they cannot be expected to carry the additional environmental costs. What this means in practice is that either the buyers or their customers/end-users have been getting the product too cheaply, at an ecologically uneconomic rate.

At this point it becomes a question of price elasticity – can the buyers afford a reduced margin or are customers willing to pay a higher price? If neither is possible, the conclusion has to be that the product is no longer viable as such.

Of course, it may well be that, in the absence of legal constraints, less scrupulous competitors are able to continue ignoring and externalizing the environmental costs. The business decision that has to be made under these circumstances goes to the heart of the environmental dilemma facing business – whether to continue to make profits in the knowledge that we are damaging our most vital and irreplaceable assets.

COMMUNITY PAYS

We are not aware of any examples of government subsidies being made available to rescue products or services which would become non-viable if they had to pay their full environmental costs, except in the very special cases of the coal industry and of nuclear power station decommissioning costs. Yet the UK government is one of the culprits complained of by the BCSD[4] for subsidizing road transport and industrial energy so as to disguise their full environmental cost and remove the market's incentive to tackle their impact. From this point of view, a government subsidy to clean up products judged both essential and environmentally non-viable would no doubt be seen as equally undesirable.

VALUE FOR MONEY IN THE PUBLIC SECTOR

Public sector procurement, in both national and local government, is increasingly applying environmental criteria. In so doing, it can come up against the traditional 'value for money' (VFM) imperative to satisfy the functional requirements at the lowest available cost. Often this is seen as a form of *force majeure* which prevents the purchase of a more environmentally acceptable alternative at a higher price.

Leaving aside cases where the 'greener' product can actually provide cost savings (it can happen!) or where suppliers' margins are big enough for them to drop their 'green premium' to retain the business, there are bound to be occasions when goods that meet environmental criteria are more expensive than those that do not, because they are made by less polluting but more costly processes, or from less harmful but more expensive materials. The suppliers of these goods are effectively 'internalizing' in the cost of the product the real environmental costs, which are otherwise borne by the community and the planet, here or overseas.

These are the situations where the financial policy makers, whether in the Treasury or the town hall, are having to think again. A government which is committed to 'sustainable development' on a global scale, or a public body with a declared policy of minimizing environmental damage, is in difficulties if it defines the more environmentally damaging option as better value for money, unless the additional cost of purchase exceeds the externalized environmental cost to be avoided.

A practical procedure for dealing with these cases might include the following:

1. *Obtaining information* about differences in cost and environmental performance – i.e. answering the question: What environmental benefit would the extra cost be buying?
2. *Guidelines* about the importance to be given to different factors, such as:
   - resource costs (renewable/non-renewable, recycled/recyclable)
   - life-cycle costs (environmental costs related to extraction, manufacture, transport, disposal)
   - environmental costs in use (emissions, waste, use of energy, water, materials)
   - supplier's environmental policy, record, management system (BS 7750).
3. *Designated level and accountability* for taking decisions on environmental versus VFM dilemmas.

Quality and specification – A marketing issue

QUALITY – AN ELUSIVE CONCEPT?

There is an increasing tendency for environmental attributes to be perceived as an important part of total quality. For instance, stationery, advertising and publications using recycled paper have gone from the image of an inferior substitute for the virgin white alternative to being widely perceived as evidence of coming from a caring, quality company.

The very argument of some producers that 'green is a luxury we can't afford' helps to create an upmarket impression. This trend will be reinforced when ecolabels start appearing on some of the more expensive imported products from countries where quality and environmental performance have become practically synonymous.

SPECIFYING GREEN PERFORMANCE

At the same time it is becoming more usual for purchasers to include environmental aspects in their specification as a way of strengthening their green marketing. These may be features of the product itself, such as its energy efficiency or its direct environmental impact, or they may relate to earlier stages of the life cycle, such as non-use of ozone depletants and other environmental impacts of production, right back to the extraction of the original raw materials. (Example 9.2 describes a supplier policy based on Life-Cycle Analysis.)

Attempts have been made to specify sustainable production methods for wood, especially hardwoods, until it was realized that reliable evidence of this cannot be produced at present. The WWF '1995 Initiative', which is supported by a group of leading companies, then set about creating a system of verification for sustainable production of timber.

Other areas likely to be specified will include biodegradability and/or recyclability of component parts or substances and even packaging. Under pressure from certain governments, manufacturers of automobiles, computers and other consumer durables are moving towards the completely recyclable product.

In due course, supplier specifications may also include certification of the supplier's environmental management system to BS 7750. All these options will have to be considered, particularly in tendering situations subject to the EU Directive which could make it unlawful to discriminate against environmentally unfriendlier suppliers if the desired green criteria had not been explicitly included in the original specification.

Scott Ltd – Analysing suppliers' environmental impacts

Scott Ltd, UK subsidiary of the US Scott Paper Company, has established a formal environmental policy, committing it to understand the life-cycle impact of its products, to minimize its own environmental impacts and to incorporate environmental standards in its supplier selection process.

To achieve the third of these objectives, suppliers of Kraft pulp were asked to complete a questionnaire giving data about their emissions to land, air and water per tonne of pulp. Different releases were weighted by consulting 'opinion formers', resulting in the main emphasis being given to releases of CO_2, followed by organochlorine discharges suspended solids, biological oxygen demand (BOD) and acid rain emissions.

It was found that the range of emissions was very wide, with, e.g., a factor of 17 between the highest and the lowest CO_2 emissions. The energy sources used have a lot to do with this, but the suppliers who performed worst on one type of emission tended to be worst on the others as well.

Suppliers were given feedback of their own positioning in relation to the overall range of results to enable them to review their own processes. The results of the analysis were used in the following year's supplier selection process, in combination with other criteria, resulting in the worst 10 per cent of suppliers being dropped.

Example 9.2 Scott Ltd

Working with suppliers: selection, development and partnership

One aim of purchasing policy has always been to develop effective and reliable sources of supply. This means cultivating long-term relationships with selected suppliers who become familiar with the buyer's requirements. In the context of environmental strategy it also means ensuring that preferred suppliers can meet the company's environmental standards and, if necessary, helping them to do so.

The first requirement is for information about products, ingredients and suppliers' environmental credentials. While large buyers like Scott, BT or IBM do not seem to experience much difficulty in obtaining the information they require thanks to their size and commercial muscle, smaller organizations can still find information a major hurdle in implementing an environmental policy.

In the UK local authorities are overcoming the problem by setting up buying consortia and compiling databases. Ecolabelling will help if the scheme becomes fully effective, but so will the spread of green purchasing policies throughout business, as producers find that provision of environmental information is a condition of trading.

Some of the pioneers of green purchasing have preceded their requests for supplier information by a major information effort of their own. This may include circulation of both their overall environmental policy and their purchasing criteria to all suppliers, with an explanation of their reasons and a request for the supplier's help, possibly followed by special meetings, presentations and visits.

Help for suppliers in gearing themselves up to meeting the buyer's requirements may include guidance notes, fact sheets and explanations of environmental issues (B&Q), self-assessment questionnaires (IBM), training seminars for suppliers' staff (Boots).

These will be followed by requests for detailed information about the supplier's products and business, usually in the form of a questionnaire. Some buyers classify suppliers into categories, ranging from peripheral off-the-shelf purchases to providers of the most highly sensitive key components, and require different amounts of information and degrees of verification accordingly. Verification may include site visits, use of consultants and independent audit reports.

Questionnaires can be used creatively for much more than one-way information gathering. Scott fed back information about individual suppliers' comparative performance on the main environmental measures being used, in relation to the whole data set. This kind of action not only helps suppliers to improve their performance individually, but in the long run is likely to lead to a general raising of standards.

Where the information-gathering stage leads to the identification of problems in meeting environmental requirements, it is possible to invoke the latest ideas about buyer–supplier 'partnership', leading to the buyer and supplier working together towards achieving sustainable alternatives to existing products and technologies, with both partners raising their competitive capability in the process.

Integration with other departments

For procurement to maximize its contribution to the company it has to integrate its operations, often responsible for a large proportion of total turnover, with those of the main departments and functions whose continued effectiveness it serves. This requirement is especially important for environmental strategy, where the effects of a green procurement strategy will only bear fruit in so far as they support and are supported by the strategies of operations, marketing, R and D and other functions.

In particular, the effectiveness of environmental management itself will be strongly dependent on maintaining close links and collaboration with procurement and logistics activities. This becomes obvious as soon as environmental performance is not seen as restricted to the company's internal operations but is looked at in terms of the entire product life cycle.

9.2 Logistics and the environment

In some companies Logistics is organizationally linked to Purchasing within the overarching function of materials management. But even if the two are organizationally quite separate, they are liable to overlap in practice as regards their environmental impact.

On the one side, the selection of suppliers of raw materials and components (and the supply chain of which they are a part) gives rise to differential transport requirements, which in turn affect both financial and environmental costs.

On the other side, the procurement decisions which govern purchase or leasing of vehicles, or alternatively the selection of haulage contractors, as well as the procurement of packaging and of waste disposal services, can be very environmentally sensitive. They are decisions which impact directly on fuel usage, and hence on a major source of pollution and CO_2, and indirectly on all the environmental impacts of the motor industry and all the forms of noise, congestion, contamination and waste associated with motor vehicles on the road and after they are scrapped.

Attitude to logistics environmental strategy

According to a recent estimate, transportation now accounts for one third of all energy used in industrial countries. In spite of both the cost and environmental implications of this, within industry logistics is not generally seen as a key area for environmental strategy so far, although the operations falling under this heading make significant and growing contributions to atmospheric CO_2, pollution and waste, to be added to those made by power generation, manufacturing and domestic consumers.

Why is this? The reason may be that there is a sort of defeatism in the air – there seems to be nothing much the individual firm can do that would make a great deal of difference. The impression of many managers is that, apart from the obvious measures that they would be taking in any case in the interests of economy, they are stuck with a distribution system which is very difficult to change because of the infrastructure on which it is based.

The global BCSD and Britain's ACBE, both made up of the heads of leading companies, agree about the critical importance of reducing CO_2 emissions by greatly increasing the efficiency of road transport. They see this as a joint task for industry, who need to review hallowed practices in both freight transport *and company car allocation*, and government, who should be providing tax incentives for fuel efficiency and increasing fuel taxes so that prices begin to reflect the true environmental and socio-economic costs of fuel usage. Such action, in spite of current reluctance, is bound to spread as growing climatic disturbances increase pressures on and by governments.

It is also dangerous to equate environmental improvements through fuel efficiency with cost savings alone, because it carries the damaging implication that, if the cost saving is not worth having, then neither is the environmental one. A company which is ready to undertake all sorts of other actions to reduce its environmental impacts should at least regard any possible improvement in energy efficiency as worth while if financially it breaks even.

Logistics and the greenhouse effect

Carbon dioxide (CO_2) is the major product of burning fossil fuels, including oil. Though harmless in itself, it joins the 'greenhouse layer' of existing CO_2 and other gases which prevent solar heat from dispersing. While the full mechanisms of temperature and climate change are still not completely understood, there is a strong presumption that by continuing to discharge CO_2 and other greenhouse gases at the present (increasing) rate we are likely to precipitate rises in global temperature, leading to major climate changes and widespread flooding of low-lying areas. In accordance with the now accepted 'precautionary principle', governments have agreed to stabilize CO_2 and other greenhouse emissions in the Convention on Climate Change endorsed by the Rio Conference. Over and above this, there is strong scientific opinion that stabilization will not be enough and that considerable reductions are necessary. But given the present rising trends, even stabilization (currently set at 1990 levels) is impossible without considerable reductions of greenhouse gas emissions in a number of areas. Logistics is one of these.

In 1991 transport in the UK was responsible for 30 million tonnes of CO_2 emissions, 19 per cent of the UK total, and had increased by a quarter in the previous six years.[5] The UK government's main responses so far have been the decisions to increase the price of fuel by progressive tax increases and to announce its intention to launch a programme of road pricing. As the climate change issue becomes more critical other measures may well follow, but they are all likely to do no more than underline what industry already knows: that it can save money and improve the environment by increasing fuel efficiency and reducing distance travelled.

'Cooptimizing' environmental and economic costs will increasingly mean giving additional weight to fuel efficiency/economy in vehicle purchase and in the management of transport operations. As regards the latter, research and leading-edge experience show that there is a lot of scope for improvement even in the existing context. Driver training in 'defensive' fuel-efficient driving can produce savings of 10 to 15 per cent, but needs to be combined with a management system which includes continuous monitoring and incentives.[6] In a recent survey of members of the UK's Institute of Logistics, companies were asked about their adoption of 10 types of fuel efficiency-related policy. More than 90 per cent had policies for preventative maintenance, and more than 80 per cent for correct vehicle/load selection and route rationalization. Two-thirds saved fuel by backloading (ensuring that at least some of their vehicles did not return empty – though only 28 per cent managed to achieve this all the time), by the use of aerodynamic body kits and by driver training (64 per cent for some and 46 per cent for all drivers). Other policies for saving fuel included speed limiters, reducing unladen weight, fuel additives and driver incentives.[7]

One important decision where environmental and financial considerations may be in conflict is the question of 'modal choice', i.e. whether to use road or rail. But although in theory it is taken as read that rail, if feasible, is usually the more environmentally-friendly mode of transportation, and although in recent years much thought has gone into developing 'intermodal' or 'combined' road/rail solutions,[8] the above-mentioned survey found that fewer than 20 per cent of companies responding even have *a process for considering* anything other than road transport.

It should be emphasized that, while the decision to use road transport in any particular situation may be a matter of economics and practicality, whether to have a process for deciding is clearly a strategic issue. Failure to consider the possibility of using less environmentally damaging forms of transport, not only under an existing set of circumstances but also as a result of intervention and influence by the organization, is equivalent to ignoring other forms of pollution because the legislators have not caught up with them yet. Here, as elsewhere, the economics will change once the customers take an interest. Indeed, the UK Institute of Logistics' handbook *Logistics and the Environment* quotes a study which shows 'combined transport' to be highly competitive, with a 15 to 18 per cent cost advantage over road alone.[9]

The other sensitive area of logistics is, of course, warehousing, and both the location and construction of depots can have an impact on CO_2 emissions by affecting energy usage.

Logistics and pollution

Road transport is acknowledged to be one of the major sources of pollution worldwide, and particularly in the developed world. Exhaust gases contain a whole cocktail of poisons, including:

- lead (where leaded petrol is still used)
- the deadly carbon monoxide
- nitrogen oxides (NO_x) which join sulphur dioxide (SO_2) from power stations to create the acid rain that kills forests and poisons lakes and rivers, as well as causing multiple forms of damage to human health
- unburnt hydrocarbons (irritants as well as suspected carcinogens)
- the small carbon particles (black smoke from diesel engines) which bind further hydrocarbons that may be cancer-forming.

In petrol engines a three-way catalyst is a necessity to meet the emission limits set by US and European legislation, but does not eliminate the gases completely or consistently through the warm-up cycle. Diesel engines, which produce significantly lower volumes of CO_2, hydrocarbons and carbon monoxide than petrol engines with catalytic converters, produce higher NO_x emissions, which have recently come under increasing scrutiny as a source of human ill-health and acid rain. All in all, there is at present no 'fix' to eliminate air pollution from vehicles, and all the existing palliatives are likely to be overtaken by the continuing growth of traffic volume.

Hence the strategic question is the same for pollution as it is for global warming (and energy cost saving) – how can road traffic be stabilized and then reduced? It should also be remembered that road freight is by no means the whole, or even necessarily the major part, of business's contribution to the growing volume of environmental damage by transport. *Company car policies*, usually administered quite separately from freight transport, are still one of the major ways in which business is compounding the problem and through which it could contribute to the solution. The fact that these are often administered by personnel departments who, in our experience, are among the least environmentally informed people in many companies (with the possible exception of accountants) is a symptom of the broad failure to adopt a strategic approach to the firm's environmental impacts.

Logistics and the ozone layer

Many people believe that the fight against the destruction of the atmospheric ozone layer has been effectively won, by widespread decisions to phase out the main ozone depleters, starting with the group of chemicals known as CFCs. This confidence is unfortunately misplaced. It is true that CFCs have largely (not entirely) disappeared from aerosol cans and are being widely replaced as solvents; however, the move out of CFC-based refrigeration is just beginning, and most of the refrigerated transport and warehousing which constitute the special contribution of logistics to this problem are still based on the CFCs or their replacements, the HCFCs, which are less powerful ozone depletants but slower to dissipate and contribute to the greenhouse effect. (It has been agreed internationally that HCFCs will also be phased out.)

It is now known that the present ozone 'hole' is bound to increase for decades after all releases of ozone depletants have been stopped, with the incalculable effects of ultraviolet radiation on human and animal eyesight, on skin cancer, on the yield of food crops which are already failing to keep up with population growth, and on some of the mechanisms which could otherwise slow down global warming effects. For the earth, for humans and other living beings, this is undoubtedly a 'strategic' (i.e. long-term survival) issue, while business seems still to be treating it with only partial seriousness. Companies are happy to take credit for what is so far a very patchy 'phase-out', subject to commercial constraints and the availability of 'drop-in replacements' within existing technologies. However, few seem willing to radically rethink their technology strategies in response to the desperate urgency of a crisis which can no longer be averted but can still be prevented from becoming even worse.

For the present, for all its 'consensus' on the subject, world business is continuing to pump ozone depleters into the atmosphere. The conviction that short-term financial returns are more important than long-term security has hardly been dented.

Logistics and waste

The various logistics functions in a company are likely to make an important contribution to the overall effectiveness of its waste management, particularly in the following three areas:

- secondary packaging

– the wastes occurring in vehicle and warehouse operations
– the disposal of vehicles at the end of their life.

Some of the possible actions to 'eliminate, reduce, reuse or recycle' waste may turn out to be highly remunerative, while in other areas it may be necessary to work quite hard to reduce noxious impacts of waste for what may only be a moderate return. Once again, the strategic question to be addressed is not 'How much can we save?' but 'How can we minimize environmental damage in the most cost-effective way?', remembering that the accumulation of unrecovered waste, as such, is damaging in the long term.

Secondary packaging is an area where attractive savings have been, and continue to be, made. As logistics is mainly concerned with the transport (and intermediate storage) of materials and products between companies, there is much scope for agreement between dispatcher and receiver to replace throw-away packaging by reusable forms, to increase the number of 'trips' of reusable containers and to set up recycling arrangements for shrink-wrap and other plastic materials. Environmentally-conscious firms set up arrangements with their suppliers and customers to cut out unnecessary packaging and find reusable or recyclable forms of product protection.

When it comes to *vehicle operation*, the most important forms of waste are tyres, batteries and engine oil. In the UK alone 30 million waste tyres (including about 2 million from goods vehicles)[10] have to be disposed of each year. In the past the vast majority of these unwanted tyres went to landfill or were disposed of illegally ('accidental' fires on dumps whose owners had disappeared). The EU has plans to recover all tyres by retreading, recycling or as fuel by the year 2000, when it should no longer be possible to dispose of either whole or shredded tyres to landfill.[11] In the meantime arrangements for voluntary recovery by all of these means are progressing rapidly, and an environmental strategy for logistics should include ongoing evaluation of a growing range of possibilities. Similar considerations apply to batteries and lubricant oil, for both of which recycling schemes exist.

Disposal of vehicles, lift-trucks and other equipment at the end of their working life is becoming a significant part of environmental strategy. Before selling a vehicle for scrap the waste-management hierarchy of 'reuse, repair/remanufacture, recycle' should be considered. Many vehicles can have their life significantly extended by replacing worn parts, manufacturers are under increasing pressure to take back vehicles in order to close the loop by a combination of remanufacture and recycling, and environmentally concerned customers can negotiate these options at the time of purchase. Managers should remember that the amount of recycling undertaken by scrap merchants may be very variable and fall far short of what is either possible or desirable.

Elements of an environmental strategy for logistics

OVERALL OBJECTIVE

It is important to be clear about the aim of such a strategy. Primarily it is to contribute to the organization's total environmental performance, not to discover new ways of saving money, although these are likely to be an incidental benefit.

REDUCING AND REDEPLOYING

The overall need for transport can be reduced by depot siting, backloading and distribution strategies (e.g. moving products between dealers), and redeployed from more to less environmentally damaging modes, particularly from road to rail. This may involve ongoing procedures for evaluating the possibilities of redeployment, as well as proactive efforts to influence carriers' rates and policies and to make an input to government transport planning.

FUEL EFFICIENCY

As indicated above, there is a wide range of paths to increased fuel efficiency which, taken together, can make a substantial difference both to CO_2 and other emissions and to the company's costs. Reduction of these emissions should be one of the main aims of environmental strategy.

REFRIGERATION

Rapid phase-out of ozone-depleting refrigerants should be treated as a matter of urgent priority.

WASTE REDUCTION

A systematic programme to enable the reuse, remanufacture or recycling of vehicles, equipment, packaging waste, tyres, batteries and lubricants is the other important component of an environmental strategy for logistics.

References and further reading

1. D. W. Dobler, D. N. Burt, and L. Lee Jr., *Purchasing and Materials Management* (5th edn), McGraw-Hill, 1990, show these functions as nesting within each other in order of increasing generality.
2. Adapted from Dobler et al, op. cit.
3. The quality and performance revolution sparked off by Japanese competition and investment, described, e.g., for the motor industry by R. Lamming, *Beyond Partnership: Strategies for Innovation and Lean Supply*, Prentice-Hall, 1993. The term 'lean' is applied to a production system that 'operates with a bare minimum of everything: labour, materials, organizational complexity, space, etc.' It leads to a requirement for maximum supplier input to process and product development, described as going 'beyond partnership'.
4. S. Schmidheiny (ed.) with the Business Council for Sustainable Development, *Changing Course*, MIT Press, 1992.
5. Department of the Environment, quoted in ENDS, May 1993, page 4.
6. S. Bedford, 'McKelvie and Heriot-Watt University', *EnviroEconomy*, 1993.
7. *Going Green*, PE-International Logistics Consulting Services, 1993.
8. Two such systems, the Road-Railer and Piggyback Transport, are described in the Institute of Logistics' handbook *Logistics and the Environment*, Vol. 1, 5-46–5-47.
9. Ibid., Vol. 1, 5-44. Earlier in the same chapter (5-43) it is pointed out that the EU has been supporting combined transport in various ways.
10. *Logistics and the Environment*, Vol. 1, 4-54.
11. ENDS Report 220, pages 35 and 226, page 39.

10 Human resources strategy

1. It is time human resource specialists became more involved in environmental strategy. Its success depends on strategic capability, management development and mobilizing staff support, all of which are part of HR expertise.
2. Human resource issues involved in environmental management include:
   - Recruitment and selection
   - Discipline and control
   - Motivation and reward
   - Performance appraisal
   - Communication, involvement and ownership
   - Change management.
3. Most environmental training is at present being done by environmental rather than training staff, but could benefit from professional training expertise. Included are:
   - Awareness training at all levels
   - Job-related and technical training
   - Training in management systems and environmental audit, induction training.
4. Beyond specialized environmental training, there is a need, at present unrealized, for true management development to open up the creative new approaches to sustainable management that are still lacking. This calls for the involvement of HR development professionals. Possible approaches to meeting this need might include:
   - involving managers in 'cascade training'
   - projects
   - Action Learning networks

- self-managed learning/education
- function-related environmental training.

5. We believe that it is important for HR professionals to become involved in all the human aspects of environmental management in order to fully release the capacity of business organizations to solve the environmental issues we face.

Ultimately, sustainable development will result from the billions of choices individuals make every day. It will happen when people are committed to making it happen. And the time to make it happen has arrived. (Business Council for Sustainable Development, *Changing Course.)*

10.1 What does human resource management have to do with the environment?

Very little? At first sight one might be forgiven for replying 'Not much', to judge from the apparent lack of interest among HR professionals. The environment is rarely mentioned in personnel journals. Courses on environmental training design are mainly attended by environmental specialists, rarely by professional trainers.[1] This is because most of the environmental training that is done within companies is undertaken by the environmental department, rather than by Human Resource Development (HRD) or Training. With a few notable exceptions, it is difficult to find companies which include environmental targets or performance in their performance appraisal systems.

Or a great deal? Irrespective of how much, or little, interest HR professionals are showing in environmental management, there are plenty of reasons why they should be interested.

Strategic capability

If a company is serious in believing that its future depends on addressing environmental issues and achieving sustainability in its operations, then it faces a major problem – how to make it happen? Strategy without strategic capability is fated to remain an empty aspiration, a 'might-have-been': another way of saying that strategy depends for its success on the capability of people to make it happen.

How to become sustainable is not something you can read up in a book – it has to be invented. For this, the organization needs people who have a clear understanding of the issues and the business, and who possess the knowledge, skill, motivation and creativity to discover innovative solutions that work. This means that recruitment, selection, appraisal, training and development – the stock-in-trade of Human Resource Management (HRM) – are essential building blocks for any genuine attempts to move towards a sustainable business.

Management development

Environmental management is not simply a body of specialist knowledge, but a new approach to management, involving a revolutionary new assumption, which is frankly at odds with the traditional views on which most management development is based. We have described this change in Chapter 3. It consists in moving from the restricted focus of 'bounded rationality' towards a broader 'whole system' perspective.

Conventional management theory focuses decision making on the pay-off from a bounded and partial system, seeing the surrounding world merely as a source of resources and constraints. By contrast what we have described as *ecological management* (Chapter 3) requires the manager to take responsibility not only for the effectiveness, success and survival of the commercial entity in his or her charge, but also for its impacts on the larger system which is partially unknown, for practical purposes unbounded, and which includes unknown and non-commercial, perhaps even non-human, stakeholders.

These are not dogmatic assertions of a new 'truth', but simply an attempt to point out the inherent implications of the view that environmental management means, in a paraphrase of Fritz Schumacher, 'managing as if the environment matters'. The reverberations of this view for HRD are potentially enormous. It could, and should, mean rethinking the whole approach to management development, involving perhaps a partnership between education and business, with HRD/HRM providing the linkpin between the two.

Mobilizing support

Edgar S. Woolard, who as CEO of Du Pont described himself as its 'Chief Environmental Officer', saw the environment as the strongest force for 'coalescing the organization about a common purpose'.[2] Companies are full of 'closet greens' who feel frustrated if the company plays down green issues or fails to make use of their knowledge and enthusiasm, but who could be enrolled to provide the driving energy and practical realization for an ambitious environmental strategy.

Mobilizing people towards a common purpose is traditionally one of the functions of HR departments and their predecessors in personnel and employee relations. It is time they woke up to the powerful dynamic which a drive for sustainability can represent.

10.2 What are the human resource issues?

When you focus on what day-to-day environmental management is really about, it becomes plain that, side by side with the management systems (the audits, the effects registers, the impact assessments) and the technical means (the bunds, the scrubbers, the emission monitors), the path from environmental protection to sustainable development is punctuated with issues of 'people management'.

Training and development are obviously the key HR issues for all companies trying to take up the environmental challenge, and these are dealt with more fully below. But quite apart from this obvious need, there is a whole series of people-related concerns which come up as soon as a company starts to think seriously about its impact on the environment.

Recruitment and selection

As companies wake up to the need for managing their environmental impacts, they look around for people with the necessary knowledge and experience. The demand for men and women who combine relevant knowledge with environmental management experience is likely to grow more quickly than the supply, particularly as the consultancies cream off some of the best people becoming available.

Human resource managers will be more successful in selecting this new type of resource for their company if they seek to gain a better understanding of the issues which the people they appoint are going to have to deal with and become involved in their company's strategic thinking about the environment. Does the company require specialists able to apply the quick fix without disrupting ongoing plans? Or is the company set on a path of environmental and managerial innovation which requires the type of person described by the Prospect Centre as 'capable of thinking futuristically, and free of the restraints imposed by corporate norms'?[3]

Discipline and control

Simple questions of discipline cannot be avoided, when a company can be heavily fined and individual managers may even go to prison for failing to prevent serious pollution incidents. But some employees, and even managers, will resent and resist environmental policies and requirements as an interference with established working patterns or an additional burden.

In one factory a damaged machine as big as a car was found in a waste paper skip. In another, unsupervised workers on night shift saved time by tipping toxic waste into the local river. The first infringement interfered with the waste recovery system but did not cause direct environmental harm; the second, though less striking, laid the company open to criminal prosecution.

Whether environmental rules are enforced through strict disciplinary procedures or dealt with in more involving and participative ways or through better training and awareness are questions of HR policy and relate to the ways in which employee relations are structured in the company.

Motivation and reward

How to motivate staff towards good environmental practice and find ways of publicizing and rewarding exceptional performance are questions to which HRM will know many of the answers. In some companies these tasks have been tackled very successfully by special communication exercises, by environmental awards for exceptional performance and ideas, even by incorporating environmental criteria in salary reviews. In other companies these questions have been answered by default, to the detriment of environmental strategy.

At Du Pont 'environmental performance has been made one of the criteria for judging the compensation of senior managers. Senior and junior managers alike are evaluated for their compliance with company policy and legal requirements, their performance on corporate goals, and the impact of self-initiated programmes ... an environmental awards scheme ... (is) designed to recognize individuals, groups, and businesses that have displayed superior environmental stewardship ...'[4]

Performance appraisal

Through their performance review and appraisal systems many companies have a very good opportunity to reinforce their environmental policies and targets and make it clear to their staff that they mean what they say. So far, however, we have had difficulty in finding many firms that have in fact included environmental improvement goals in the individual targets against which managers and other staff are reviewed. This is an area where environmental policy implementation has much to gain from the closer involvement of HR staff, who usually run these performance review systems.

Communication, involvement and ownership

Instructions, discipline and motivation are important factors, but it is doubtful whether they will be sufficient to turn around the unsustainable way in which many companies, and the economy, are operating. Finally, as the quotation at the beginning of this chapter points out, individuals have to take the right, sustainable decisions in the myriad transactions in which they are involved every day, at home and more powerfully, at work.

Companies have long ago discovered that to make key policies effective they have to be 'owned' by every member of staff. This has been demonstrated by the 'total quality' revolution, which succeeded in those organizations where everyone was involved in developing new approaches, resulting in a major change of mindset and attitude all round. Similarly, success in changing industry to make it environmentally sustainable will require full 'ownership' of the goal by all who work in the system.

There is a simple way by which people gain ownership of a decision or a policy, and that is to participate in the making of it. After all, the state of the environment and how to preserve it without shutting down our factories, our power stations, our transport systems, is everybody's problem, not just the board's. Involving people in working out what needs to be done, particularly at the level of their own job, work group, department and business unit, is the surest way of getting them to make the policy their own.

Companies like 3M and Dow Chemical, which have introduced employee programmes to generate money-saving ideas for waste reduction and pollution prevention, have released a wave of creativity and new ideas which have greatly strengthened their environmental improvement plans as well as impacting favourably on profits. Dow's programme 'is driven by a combination of senior management commitment and employee initiative .... Team effort from motivated employees has been the key to success .... One result of the ... programme is that less than 1 percent of Dow's hazardous waste now ends up in a landfill.'[5]

In those firms where HRM is still regarded, and regards itself, as a key resource for maximizing individuals' and groups' contributions to corporate goals, they could be playing a leading role in communicating the environmental messages through systems of employee participation and involvement. Example 10.1 shows how successful a participative team approach can be.

Renishaw: Environmental improvement through team working

Renishaw plc are manufacturers of innovative hi-tech metrology equipment the worldwide sales of which have grown almost five-fold over the last 10 years. Of the 700-plus employees, more than a quarter are engaged in research and development of new metrology applications and products. The company culture and structure is flexible, dynamic and resourceful, and is driven by a 'continuous improvement' programme of quality management which makes extensive use of interdepartmental teams.

In 1992 the company decided to add 'environmental concern' to its 'core values' at the same level of priority as 'care for people' and 'quality' and to set up a structure for environmental management. For this purpose the company set up a 'Green Team' as part of the Training-led team-building programme. The cross-functional team included the quality manager, a product manager, an accountant, a purchasing officer, an assembly supervisor and a graduate engineer (who incidentally was an active Friends of the Earth supporter).

The team was charged with conducting a full environmental audit and performance assessment of the main sites and to make recommendations for action. It was also asked to propose the content and format of a corporate environmental policy for top management endorsement. The Green Team carried out its brief over a five-month period and came up with an environmental policy, an implementation guide for making it happen and a set of immediate recommendations for areas of non-compliance – in waste management, toxic chemicals and use of CFCs. These products were publicized to all employees through wall displays in the canteen (used by all levels) with ample discussion opportunities.

The board accepted the proposals, endorsed the policy and established an environmental management structure, still based on the team approach. The structure involved an environmental representative at board level, a steering committee (chaired by the board representative) and an interdepartmental implementation team to action both the immediate recommendations and the further objectives, criteria and strategies defined by the steering committee.

The implementation team was 'chaired' (i.e. facilitated) by the Quality Manager who was also a member of the steering committee, and the team members received training in interpersonal and problem-solving skills. In the meantime the Strategic Planning Manager, who had taken an active part in facilitating the development of the team structure, was running a series of monthly strategy meetings for key managers and the steering committee, to present strategies for endorsement by the board.

Early outcomes of the implementation programmes included:

1. Significant capital investment to eliminate all CFC-based component cleaning systems. The new system also provided a better working environment for employees.
2. Building of improved waste storage facilities, e.g. for solvent containers.
3. Programmes to eliminate CFC-based packaging materials.
4. Environmental aims and requirements being communicated to suppliers.

Example 10.1 Renishaw

Change management

All the approaches we have discussed are contributory aspects of bringing about important changes in the organization. Whether, like many people in developmental roles (whose bread and butter it is) we regard organizational

Change is a good thing

Except

when we are doing well
when we are under pressure
when there is a boom
when there is a recession
when it gets in the way of my work
when there are more urgent things to be done
when it threatens my interests
when it costs money
when ...

Figure 10.1 Change is good, except ...

change as good in itself, or whether, like some line managers with a job to do, we wish that all change could STOP for a period to allow us to get on with the essential tasks we have had to put off because of having to cope with too much change – either way, we are bound to agree that for a firm to even take the first steps towards becoming sustainable involves real organizational change. Most people welcome change – but not now! (See Figure 10.1.)

Because of their particular concern with changing behaviour in the organization, change management has been of special interest to HR practitioners, some of whom are likely to have very relevant expertise to offer to environmental managers finding themselves in the role of 'change agents'. As well as instructions, guidelines and procedures – the staples of a line management approach to change – as well as communication, motivation and involvement, it is worth spending time on:

- asking the basic change questions:
  Who feels the pain?
  Who owns the problem?
  Who has power to act?
- building a common vision of the future
- analysing the 'force field' (i.e. the forces, or people, likely to help or hinder)
- plotting the phases of the change process
- creating alliances
- checking the implications for the 'total system'
- considering specific interventions.[6]

10.3 Environmental training

At present a lot of in-company environmental training is being carried out by environmental specialists, rather than being taken on by HR professionals. As a result, it is aimed more at getting people up to speed with existing requirements than with changing managers' approaches to their jobs. The emphasis is first on awareness of the importance of environmental issues to the company, then on legislation and regulations (unless the latter come first, or are the only thing considered important). In addition, there may be more technical training directed at specific issues of pollution control, risk reduction, waste management, impact monitoring and environmental management systems.

Awareness training

The most popular form of environmental training is probably awareness training.[7] The actual delivery of such training takes place in a variety of forms, from a lunchtime video, a managerial pep-talk, a tape-slide presentation, to more sophisticated multi-media interactive events, involving videos, talks, questionnaires, group discussions and hand-outs.

If this kind of training works on the assumption that once people are 'aware' of the environmental problems they will know what to do about them, it is unlikely to be very effective. This is why it is desirable that the design of any such session, however short, should include some form of action planning by participants, which enables them to bring the message home to their own immediate sphere of activity, whether that means their business unit, their department, their work team or just their own jobs. Some companies include discussions of what individuals can do at home and away from work – a good idea, as long as it is not used as a way to avoid stirring things in the work situation.

In designing awareness training events it is important to take account of the four different levels (global, economy, company and individual) at which the problem operates and which impact on each other. For completeness, a fifth level – the future – should be added, so that awareness can be turned into an active force for change (see Figure 10.2).

Awareness training may be organized for working groups or departments (e.g. a marketing department that is expected to change its approach to customers to support the company's environmental strategy), or individuals (e.g. a health and safety manager who has had environment added to his or her brief).

The approach needs to be adapted to suit different levels of the company. The two ends of the scale (global and future) should always be present, but between these senior management will want to focus particularly on issues at

GLOBAL LEVEL

The natural environment
The real problems: Loss of soil and forests, ozone depletion, global warming, acid rain, air/water pollution, contamination

↓

LEVEL OF THE ECONOMY

The business environment
Changing the rules: Legislation, taxes, pressure groups, customer choices, competition, investment, insurance, business opportunities

↓

COMPANY LEVEL

The management environment
What we are doing: Environmental policy, strategy, management systems, pollution prevention, waste management, recycling, audits, targets

↓

INDIVIDUAL LEVEL

The work environment
What I can do: New ideas to change products, save materials and energy, reduce waste and pollution, recycling

↓

THE FUTURE

Sustainable company and business
How we will make our business sustainable: ?????

Figure 10.2 Levels of environmental awareness

the level of the economy, middle management needs to be aware of what is happening within the company as a result of its own strategies, policies and initiatives, and operating staff will be most interested in responses at the individual level.

SENIOR MANAGEMENT

Environmental awareness training for senior managers is the most important, but can be difficult to manage. Adapting Molière's immortal words about aristocrats, 'a senior manager knows everything without having to learn anything'. In some companies the chief executive has sponsored a special conference for senior managers, with inputs from external top managers and prominent public figures, and sessions for planning the company's response.

A large chemical company ran a three-day event entitled 'Creating Our Global Future' for senior managers from R and D, Planning, Marketing and Communications. With the aid of a wide range of inputs and using creativity techniques, the participating managers looked at the impacts of major trends and issues on the company's business and made recommendations to the board about research and product development opportunities.

Another approach adopted by some companies is to hold a senior managers' workshop, facilitated by a consultant, with the aim of jointly agreeing, or reviewing, the Corporate Environmental Policy.

MIDDLE MANAGEMENT

Awareness programmes for middle managers and specialists may be intended to increase managers' awareness in preparation for raising environmental standards and meeting rising requirements, or they can usefully be tailored to specific company initiatives, such as pollution prevention and waste reduction programmes or the introduction of an environmental management system.

Exxon Chemical International, as part of the extensive 'menu' of environmental courses described below, offered a two-day awareness course 'Managing in the Environment' in the setting of a Field Studies Council centre, where lectures and group discussions are interspersed with field exercises to gain a better understanding of ecological principles. Subsequent courses combined fieldwork with issues related more directly to workplace environmental concerns.

One large transport organization decided that all its managers, supervisors and specialists should be trained in environmental awareness, but that the cost of face-to-face training would be too high. They therefore commissioned a distance learning package, complete with individual and group exercises leading to action on work related environmental issues, as a basis for a company-wide management training programme.

WORK GROUPS

ICI developed an awareness package involving a video and a set of simple questionnaires, which were completed by people in a work or site group as introduction to a led discussion of the facts and fiction about major environmental issues and what the company is doing about them.

A consumer goods company produced a video about the company's environmental policies and achievements, for use both with outside audiences, such as suppliers, customers and schools, and also for its own employees, who were shown the video at special work group meetings led by managers or supervisors.

COMPANY-WIDE PROGRAMMES

Companies that have decided to incorporate environmental care as a basic part of their business strategy sometimes decide to put all their staff through an environmental awareness programme.

After deciding to give a high priority and profile to environmental management, the Rover Group decided to follow up a very successful 'Cascade' programme on Total Quality with a similar programme on the environment. A university department with an established programme of Environmental Management courses for managers was commissioned to develop a seven-day programme incorporating all the stages of environmental awareness (Figure 10.2). A 'tutor' group of around 50 managers and specialists from all parts of the company was chosen to take part in the full programme and to cascade the material, suitably adapted and condensed, through a second 'tier' of managers to the whole of the workforce.

Job-related and technical training

Large companies are making arrangements to provide a range of technical training programmes to people such as engineers, technicians, designers, site managers, environmental representatives and 'champions', who in future will need an understanding of environmental effects to do their jobs. Such programmes may be organized quite independently of the human resources function by special technical training departments.

At Exxon Chemical International, environmental training is part of an embracing SHE (Safety, Health and Environment) training programme which coordinates training available from a number of departments and sites of the company. The approach is an 'open market' one, with demand being driven by the company's Operations Integrity Management Practices (OIMP), a set of rules for safety, health and environment which include practices for pollution prevention, regulatory compliance and product stewardship. All managers are expected to achieve the objectives set out in OIMP, and in order to do so are offered a wide selection of short courses and modules, ranging from overview courses like:

Managing in the Environment (see above)
Environmental Control on Site
Basic Environmental Engineering
Emissions and Waste Reduction Technology

to some very focused courses on such topics as:

Product Life-Cycle Analysis
Air Emissions Estimating
Air Dispersion Modelling
Environmental Testing
Environmental Impacts of Process Waste
etc.

Courses are scheduled according to demand, and can be tailored to the needs of individual sites.

All such technical training programmes have an opportunity to go further than a simple 'how to' approach by enlisting the job knowledge and cross-fertilization between participants from different parts of a company, making the course into a potential 'laboratory' for developing the company's practices and operating standards, and enriching individual understanding and confidence. Whether such opportunities are grasped does not necessarily depend on whether HRD experience is brought to bear in the design of programmes, but at least the possibility of this kind of added value makes the case for some form of HR involvement in their planning.

Management systems and environmental audit

Because of increasing interest in the new Environmental Management Systems Standard (BS 7750), the EU's EMAS Regulation and the related requirement for environmental reviews and audits, some companies are running internal training courses for prospective environmental auditors or using one or other of the external courses available.

Induction training

Reference to the company's environmental policies and practices is often included in induction courses for new employees. There is an opportunity here for some real standard-setting, by conveying to newcomers not only *that* environmental protection matters to the company, but *how much* it matters, and precisely how the newcomers can contribute.

10.4 Management development approaches

Training programmes like those described above are performing a vital function, and often doing it very well. What they cannot be expected to do is to bring about any real integration of the environment into management thinking and practice. The message that most environmental training conveys, rightly, is that the environment is a new constraint for managers and others to take on board as they go about their normal jobs, a constraint like the budget, like safety, like the law. This is an important message, but it is not the only one that managers need to hear.

The other, and more difficult, message to get across is that sustainable development is going to depend on reinventing management and the company – that you, the managers of this business, have to work with people in other businesses, and people outside business altogether, to develop the new ways of providing the products and services people want, and the profits the company needs to survive and thrive, without damaging the natural systems and resources on which we all depend.

Because, by and large, HRD practitioners have not seen the environment as part of their remit, environmental management development is one of the great missing terms in the sustainability equation. Both industrialists and environmentalists largely see sustainability as a problem of technology and political will[8] – the idea that it might depend just as much on developing new ways of managing is hardly considered.

Development, unlike training, does not move towards a clearly specifiable goal. We know what managing for sustainable business will not look like – i.e. not like what is happening today – but we still have to work out what it will be like. This cannot be done by academics in their studies; it needs the commitment and willingness to learn of managers grappling with changing situations in an open frame of mind. This is what real management development will always be about – creating the managerial resources for tomorrow, not producing replicas of today's successful managers.

How can this be achieved? What examples can we draw on? There seem to be a number of approaches which go beyond simple training and have the potential for producing real development.

Cascade training

A cascade project, like the Rover one described above, goes further than simple training by involving managers in the delivery. It is well known that the best way to learn about anything is to teach it, but more than this, in adapting the messages to their own business units, departments, work groups, the managers are starting the process of pinpointing the exact relevance of the issues to their own management situation and integrating them into the whole process.

Projects

The use of practical projects in the work situation is a well-known medium for management development: by applying the method to some of the many real environmental problems which organizations face, the organization gets triple value – solutions, experience that can be used elsewhere and development of individual managerial capacity to cope with future challenges.

Action learning networks

'Action Learning' is a method of management development invented by Professor Reg Revans about 30 years ago and is well known to HRD people. It involves a group of managers from the same or different organizations, each working on a real management problem, meeting at regular intervals with a skilled facilitator and tutor and learning from each other's experience on their current tasks. This can be a very valuable complement to the project method of management development.

Self-managed learning/education

Management Self-Development (development both of and by self) is another favoured HRD approach. A range of distance learning packages on the environment are now becoming available, and an increasing number of institutions are offering part-time degrees and diplomas in environmental management, which usually provide opportunities for a combination of tuition, private study, project work and action learning, like group support.

Function-related training

The provision of environmental courses for non-environmental managers – for marketers, R and D, purchasing and logistics managers, financial managers, corporate strategists, even management developers – is growing. While these can be regarded as straightforward 'how to' training, they potentially go much further than this by inviting participants to make new connections between the environment and their current functional management role, to develop their perception of their job as a manager and to broaden their managerial capability by integrating into it the additional dimension of sustainability thinking.

10.5 Environmental training and development – who does it?

At present environmental training is largely the preserve of the environmental specialists themselves, but in some companies a new breed of environmental trainer is emerging.[9] In others a collaboration between training/HRD and environmental management is beginning to emerge.

In the not so distant future we believe that human resource managers will have to take up the challenge of equipping their organizations with the human capability to achieve sustainable development, or risk being sidelined to a purely administrative role when sustainability becomes 'the only game in town'. Now is the time to start addressing the environmental issues and to put their other learning programmes into the context of what is happening to our life-support systems out there, in order to prepare the organization for the fundamental and pervasive changes to come.

References and further reading

1. At least, that has been the experience of one of the authors, over four years of running an environmental seminar programme at Brunel University.
2. S. Schmidheiny, *Changing Course*, MIT, 1992, page 195.
3. *New Capabilities for the Greener Organisation*, The Prospect Centre, 1990.
4. Schmidheiny, op. cit., page 195.
5. Ibid., pages 266–7.
6. See also Colin Hutchinson, *Vitality and Renewal*, Adamantine, 1995.
7. Awareness training was by far the most frequently mentioned form of training when managers of 200 UK firms taking part in the Pilot Study for the British Standard for Environmental Management Systems, BS 7750, were surveyed by the 'cross-sectoral training group' set up as part of the study.
8. The exceptions are 'dark green' environmentalists who do not believe that a capitalist economy can ever be sustainable, and the 'dark blue' economists who believe that there are no problems which a free market cannot solve unaided.
9. ICI has a specialized Environmental Training Group, The Body Shop has an Environmental Trainer who coordinates self-managed training of the environmental advisers in each department (who also take part in environmental audits) in a sort of continuous cascade.

11 Communications and public affairs strategies

1. Integrity and accuracy of communications is essential in the environmental arena where many of the recipients are suspicious of PR and well equipped to test environmental claims.
2. The need to communicate with a new and diverse group of stakeholders emphasizes the importance of listening as well as speaking, and of finding appropriate communication channels. A stakeholder communications strategy should consider a range of two-way media.
3. A formal corporate environmental policy is the first and basic step in giving legitimacy to environmental communications.
4. It is vital to be able to support the validity of all environmental claims. Two useful codes provide guidance on the types of claim and the kinds of language and imagery to be avoided in environmental communications. External validation can be obtained through:
   - ecolabelling
   - strategic alliances with, e.g., environmental charities
   - life-cycle assessments
   - independent verification of environmental audits.
5. Extreme care is needed when managing the total communication mix through which the business projects its environmental message. Given the range and diversity of media used, it is important to ensure consistency across a spectrum which might embrace:
   - Brand management and communication
   - Packaging
   - Product and corporate advertising

- Sales promotion and sponsorship
- Public relations
- Charitable giving
- Crisis communications
- Environmental awards
- Displays on corporate space (vehicles, shops, offices)
- Corporate reporting.

6. In communications as elsewhere, strategy is a matter of choice. Under the following headings we describe a number of extant strategies:
   - Defensive/damage limitation
   - Reactive-/Consumer-focused
   - Environmental competitive advantage
   - Environment as profile raiser
   - Environment as a communication platform
   - Leadership/Customer education.

For many companies the evolution of corporate environmental policy was fashioned above all by the demands of, and need for, external communication. Yet in this key area much energy is expended to create a good impression of the company, only for it to be cancelled out later by other attitudes and statements. Conflicting messages arise because of the absence of a comprehensive strategy for the company's environmental performance as well as all public affairs communication. In this chapter our interpretation of public affairs covers the widest possible sense of public: 'of, or concerning, the people as a whole'.

11.1 Strategy, communication messages and integrity

Corporate communication strategies are seen by cynics as a means of manipulating audiences by creating an idealized picture of the company irrespective of the real facts. Some critics have an enormous distaste for Public Relations as a modern business tool, seeing it as gaining an unfair advantage by deploying riches beyond the means of others. The competent selection and sensitive management of corporate communications, as part of the company's corporate environmental strategy, is therefore critical. Without doubt this is a high-risk, but nevertheless manageable arena, which can determine whether outsiders have confidence in the company's integrity, and ability to make significant improvements in environmental performance.

Without integrity in intent and the ability to deliver positive practical outcomes, no communication expert or medium can hope to deliver results. Any claims or consumer propositions must be supportable. People *en masse* are no more convinced of words and aspirations than the individual, if all there is to rely on is talk and more talk. This is particularly important in the environmental sphere, as one of the greatest achievements of the environmental movement is that it has created a series of mechanisms and organizations by which individuals can test the claims made by companies and gauge the accuracy of their statements. Hence the large number of 'green' guides appearing from the late 1980s onwards, covering everything from supermarkets to travel companies, moving logically towards profiling companies themselves against a selection of common ethical and environmental criteria. For examples, see the company profiles in *Changing Corporate Values*.[1]

Without a commitment to present the case with accuracy and no underhand motives, don't be surprised if a dedicated amateur armed with no more than a PC and immense tenacity finds you out! To quote the US leadership guru Stephen R. Covey, 'Learn to walk your talk'.[2]

11.2 Communication channels with stakeholders

Stakeholder analysis and audience selection

Effective communication is based on knowledge of one's audiences, their concerns and needs. The acceptance of the rights of, and responsibilities towards, the company's stakeholders (Chapter 3) has major implications for communication strategy. Attempting to meet stakeholders' legitimate concerns, through the provision and acceptance of information, brings a profound change to both the scope and tenor of communication work.

No longer is the communications manager's role one of presentation of selected messages, working solely from the company's own agenda; new skills are now required for both speaking *and* listening, often to feedback that has hitherto been suppressed or devalued. In this stakeholder context a hitherto unknown degree of honesty may also be the best policy. These changes in emphasis, audience and skill base need to be managed effectively.

The nature and concerns of stakeholders have been covered in depth in Chapter 3. Accepting the logic and legitimacy of a broad range of stakeholders' interests of necessity leads to a total revision of the communication strategy. Historically, for many organizations 'strategy' simply consisted of broadcasting lists of carefully selected product benefits or the most flattering 'angle' on corporate activity. With the advent of Stakeholder Communications, the line of exchange becomes two-way; one listens as well as broadcasts. To achieve that change, from *corporate* communication to *stakeholder* communication, requires development of both new attitudes and communication channels; allowing the voice and concerns of the stakeholder to be heard. A preliminary review of stakeholders is an essential first step.

The ability to listen to and communicate with stakeholders necessitates understanding of *their* circumstances and concerns. Table 11.1 presents some of the avenues for reaching out to these groups, allowing their input to influence and shape the overall stakeholder communications strategy. By a stakeholder communications strategy we mean the planning of a company's interaction with stakeholders and all other significant audiences, such as the media.

Table 11.1 Stakeholder access: a checklist

A. Monitoring/Data management system

Special interest groups abound together with corresponding publications; your information management system will require restructuring to accommodate these new sources of information.

B. Stakeholder groups/Surveying opinion formers

(i) Stakeholder groups
Panels of stakeholders, gathered with an open brief to solicit views on the company, its area of operation and trends related to its environmental performance, will generate unexpected perspectives on the way in which corporate actions are perceived and what directions your corporate strategy might take.

(ii) Surveys of opinion formers
The regular monitoring of selected opinion formers to solicit their views is an established practice, and regular syndicated surveys take place. The adaptation of these surveys to include environmental issues provides an 'ear to the ground' as to the penetration of these concerns among influential groups.

C. Market Research

(i) Focus groups
This classic market research technique for interviewing consumers is easily extendable to take in the views of particular stakeholders and varies little in intent from a wider stakeholder group, the difference being in approach and methodology.

(ii) Attitude surveys
Syndicated sector research, which identifies attitudes among nominated target groups, is available in most established consumer markets, and existing surveys can be stretched to include a few pertinent questions on relevant areas of interest. Some national surveys of trends, such as the British Social Attitudes Report, already highlight attitudes to the environment.

D. Awareness days

The creation of opportunities for face-to-face exchange of views is critical, and many companies have instigated Awareness Days and Open Days as a means of reaching out to stakeholders. The selection of an event to focus on environmental activities through an open invitation to visit the operation gives ample opportunities to meet individuals, representatives of relevant social organizations, neighbours and local environmental groups.

E. Feedback questionnaires

The solicitation of views on company performance through questionnaires for marketing purposes is commonplace, and their use as a means of monitoring stakeholder concerns and attitudes is likely to grow.

Armed with an improved understanding of the nature and interest of these groups, it is possible to begin the task of designing appropriate communication messages and channels.

11. 3 Corporate environmental policy (CEP)

Legitimacy

The authentic is said to 'ring true', and to create legitimacy in communication requires a demonstrable commitment to the environmental agenda. The creation of communication messages is an inexact science and one where talent and flair are highly prized, hence the huge sums paid to advertising creatives. Beyond using innovative images and bright jingles, the core of any good communication has to be the strength of core offering.

As content matters as much as the form of presentation, it is necessary to be able to substantiate the claims under consideration, to prove to the outside world beyond doubt that they have legitimacy. The basic step in this process is the development of the company's own internal environmental strategy, while the first *public* action for any organization intending to take environmental matters seriously is to develop a formal written policy regarding environmental responsibilities. This is a clear signal to everyone, both inside and outside the company, that environmental issues have a new profile.

Some companies are nervous as to how public their commitment should be, and attempt to have both public and private faces, fearing ridicule and the attentions of the pressure groups if they are unable to meet stated obligations. In most circumstances it is better to state clearly where you currently are, warts and all, if there is a genuine undertaking to improve against serious targets and such progress can be publicly substantiated.

There are various approaches to developing a policy, ranging from the creation of a document tailored to your organization to the adoption of one of the publicly quoted charters that are available. The most widely known charters include that of the Coalition for Environmentally Responsible Economies (CERES – formerly known as the Valdez Principles), the ICC's Business Charter for Sustainable Development and the Chemical Industries Association (CIA) 'Responsible Care Programme'. There are also statements by the BCSD, the European Petroleum Industry Association (EUROPIA) and the CBI's Environment Forum. (See Appendix 1 for CERES Principles and ICC's Business Charter for Sustainable Development.)

In our opinion the direct commitment contained in a well-tailored CEP provides an essential foundation for a company's environmental communications and performance, whether or not the company also subscribes to one of the published charters. An in-house policy might cover the following areas:

- Going beyond compliance with the law to take account of the best available knowledge
- International dimensions, meeting of worldwide standards
- Level of internal commitment (e.g. board-level representation)
- Attitude to disclosure of information
- Commitment to innovation, product development and selection
- Commitments to auditing, management systems, ecolabelling, life-cycle analysis, etc.
- Attitudes to energy, waste minimization and resource use
- Training and internal communication
- Stakeholder and community involvement
- Commitment to continuous improvement
- Progress towards sustainability.

Each organization will need to find a route forward that is appropriate and one that can stand up to scrutiny as required. A clear and supportable policy underpins the entire development and implementation of your stakeholder communications strategy.

Corporate disclosure requirements

We have discussed the formal obligations of companies to disclose information in Chapter 5, where the full range of reporting options is discussed in detail. Further on in Chapter 8, the question is addressed in relation to the implications of disclosure for operations management, rather than as part of a communication strategy – though, of course, the two are related.

11.4 The validity of environmental claims

Having developed policies, strategies and actions which genuinely and effectively set out to minimize the company's adverse environmental impacts, the final hurdle remains, of ensuring that the presentation of your position remains 'truthful, decent and honest' – an undertaking which has historically been a major difficulty in the environmental area.

In the late 1980s the rush of dubious claims, confused and erroneous associations, led the environmental pressure groups to fight back, the most formalized repudiation being Friends of the Earth's Green Con Awards.[3] Failures pilloried by FOE included the 'ozone-friendly lead-free petrol' (Austin Rover), energy production that failed to reveal its own environmental impacts (British Nuclear Fuels) and the request to consume more energy to combat global warming (Eastern Electricity plc).[4]

Such loose phrasing and erroneous claims have in part been due to the specifying clients' ignorance of environmental issues, as well as their communication agencies' lack of knowledge, both of the technical and consumer issues involved and what constitutes appropriate language and imagery for this highly complex area.

Standards of excellence

The confusion among clients and their agencies alike as to what constituted an appropriate use of environmental concepts in communication led to the development of a series of voluntary codes. One example is the UK Advertising Industry Code on Environmental Claims, which states that:

- Claims should not be *absolute*, unless there is convincing evidence that a product will have no adverse effect upon the environment.
- The basis of any claim should, if possible, be clearly explained.
- The cloaking of claims in extravagant language should be avoided; this will only cause consumer scepticism.
- Spurious claims should not be made.
- Advertisers must hold substantiation for *all* factual advertisement claims.

This guidance has been applied in the following circumstances.

Terms such as 'environmentally friendly' suggest an absolute quality which may be unattainable in practice, and advertisers would be wise to avoid making categorical claims implying that their products are environmentally safe. In the rare circumstances where an absolute claim is capable of substantiation, consumer understanding will be aided by making the basis of the claim clear; for example, 'environmentally friendly – wholly biodegradable' would be acceptable for a fully biodegradable product which did not harm the environment during the course of its use or disposal.

Less absolute claims – 'greener', 'friendlier' – may be acceptable where a genuine improvement in use in environmental terms can be demonstrated. Again, the basis of the claim should be explained.[5]

This self-regulatory industry standard can be contrasted with the more comprehensive aspirations of the environmental charity Media Natura, which specializes in the presentation of communication messages by environmental, development and social welfare charities (see Example 11.1).

Media Natura

Commercial Communications Charter for the Environment

This Charter establishes the following principles designed to help ensure that the pursuit of legitimate commercial communications does not harm the environment.

1. The environment provides the communication industry with a wealth of subjects and emotion. The industry obviously recognised the power and effectiveness of environmental imagery otherwise it would not be using it as it does. However, much of the imagery associated with contemporary products and services is fundamentally misleading. As a principle, the communications industry should be able to establish the legitimacy of any link it makes between the environment and products or services it promotes. Where there is a genuine link, that link should be explained if it is communicated.
2. As a principle, no association should be made between the environment and a product, where the product concerned damages the environment.
3. As a principle, companies which are responsible for causing significant damage to the environment should not be presented as caring for the environment, whether or not they take mitigating action or minor positive actions not connected with their main business.
4. As a principle, packaging should not be used as a vehicle for misleading claims of 'natural heritage' or to associate the product with positive environmental attributes it does not possess. Packaging should be entirely reusable and manufacturers and retailers should clearly state how it can be recycled. An independent body should also be set up to receive complaints and monitor the packaging industry. Misleading packaging, much the same as misleading advertising, should be banned.
5. As a principle, wild species should not be used to communicate selling ideas, if the position of the animal involved misinforms the viewer as to its natural behaviour.
6. As a principle, all direct marketing campaigns, which by their nature are wasteful, should use non-chlorine bleached and/or recycled paper. Within all direct mail campaigns there should also be included a reply-paid

envelope which can be used by the recipient if he/she wishes to be removed from the mailing list. All company/printed material (e.g. annual reviews, brochures, cards) should, wherever possible, be printed on non-chlorine bleached and/or recycled paper.

7. As a principle, companies whose products or services despoil the environment should not seek to purchase or predicate a positive environmental image by sponsorship, patronage or support to environmental organisations or educational activities rather than by action to fully remedy the damage their business creates.[6]

Example 11.1 Media Natura

Companies who take steps to review the real environmental benefits of their products, select their messages and imagery with care and seek informed as well as innovative translators, will avoid both embarrassment and time spent on damage limitation. However, even the best efforts of present environmental champions can fall foul of the authorities, as was shown by the case of Reckitt and Colman's Down to Earth range and the UK's Advertising Standards Authority (ASA) (see Example 11.2).

Conflicting views

In the UK during 1991 Reckitt and Colman launched a range of 'green' domestic cleaning products under the 'Down to Earth' brand name, after extensive research and development.[7] Positioned as a branded product for the responsible shopper, turnover had grown to £8 million in 1993. The company's history and large advertising expenditure has allowed it to gain a share in the branded 'green' detergent sector. This advertising support has not, however, been without its problems.

In 1994 the ASA ruled on a complaint made by Lever Brothers about newspaper advertisements covering claims on 'biodegradability', impacts on rivers, phosphates and citrate alternatives and the statement that the product 'isn't tested on animals'. The judgment found that Reckitt and Colman's claims implied criticism of other products which were unfounded, or were absolute when no evidence could be provided to corroborate comparison and that ingredients had been subject to animal testing even if the whole products had not.[8]

The problem of promoting environmental product benefits and of these only being communicable by raising comparisons with existing practice is one of the difficulties of establishing a new market sector. Difficulties are exacerbated by problems over life-cycle analysis methodology, despite a company's best efforts, since judgments involved in that process are difficult territory when it comes to support of claims and the need for verification. There are obviously good commercial reasons on both sides of any argument for rubbishing the methodology and copywriting skills of one's competitors, and external authorities, such as the ASA, can only adjudicate on such disputes depending on the views available from their supporting expert advisory panels.

Example 11.2 Reckitt and Colman

External validation

The need to verify the scientific background to any claim being made has long been understood in consumer marketing and communication; however, the speed with which the environmental dimension has grown has resulted in some confusion as to the appropriate verification process and a lack of consensus as to the appropriateness of the potential solutions on offer. For the company attempting to promote its environmental activities there is a case to be made for finding the means to externally validate claims, which can minimize both the time involved and the potential flak.

ECOLABELLING

The most obvious example of external approval would be the use of an externally recognized ecolabel. Waldemar Hopfenbeck in *The Green Management Revolution*[9] gives a very clear summary of the range of ecolabels available; there are currently more than 50 in operation.

Progress towards a European ecolabel has been slow; to date, only eight categories – including washing machines, dishwashers, detergents, paints and varnishes – have been awarded. After initial high hopes, delays and the length of the drawn-out process has led to grave concern by some governments, and lack of progress is seen as the consequence of bureaucratic obduracy within the Commission in Brussels, leading to some countries developing their own independent schemes (often with the long-term aim of compatibility with the EU criteria). Some environmentalists have also withdrawn their support for the ecolabel process due to concerns about the criteria being applied (anti-animal testing coalitions being most vocal), and the opposition at the time of writing is ongoing.

STRATEGIC ALLIANCES

The tactic of selecting environmental charities or interest groups for formal or informal partnerships or associations in marketing and communication campaigns is well known. Marketing of charities is becoming an increasingly sophisticated activity which seeks to mobilize the power of corporate wealth – as well as private donations – to benefit the charity. Strategic alliances are the corporate response and counterpart: a collaboration with charitable and non-governmental groups to further corporate goals.

Coupling with an established campaigning or charitable group can raise the profile of both parties in the selected area of interest, from animal-free testing for consumer goods to tree planting for paper packaging. The fact that a reputable charity will associate with a corporate organization by lending its products or service the charity's name/logo is shorthand to the consumer for the whole process of evaluation which charities go through in making these links. Coddington[10] outlines the potential for strategic alliances in environmental marketing, with the right partner lending environmental credentials, authority and creating a new audience for your message through its own membership and fund-raising channels.

The stakes in such an undertaking are high for both sides, and the choice and selection of partners is critical. A careful SWOT appraisal of the proposed partner, together with knowledge of their internal process and their objectives in the exercise, is required to draw maximum benefit from the association. Voluntary organizations are difficult to manage in their own right, especially with the constant turnover of elected representatives on administering bodies. To ensure that there is longevity built into the relationship, it is important to ensure that both sides understand the decision-making process and that the partnership is between corporate bodies and not just in the gift of an individual.

LIFE-CYCLE ASSESSMENT

In product communication there is a need to prove to regulators the validity of claims made, and in highly complex sectors, or in situations where a product benefit can be disputed by competitors, there is a need for external authorities. In the case of environmental benefits, especially where the consumers' knowledge is poor, it is necessary to have 'chapter and verse' on the environmental consequences of the product, and here the use of LCA can be extremely helpful.

Life-cycle assessment is an attempt to measure (where possible), or otherwise assess, *all* the environmental impacts or effects of a product 'from cradle to grave', i.e. from the extraction of the original raw materials through all the stages of processing, fabrication, transportation and storage, to the effects of the product in use and its final disposal (or recycling).

While there is as yet no universally agreed methodology for LCA (and particularly no agreed method of combining dissimilar impacts into a single 'score'), there are many examples of LCA being used to substantiate public pronouncements and advertising claims. (For another use of LCA, see Chapter 7.)

Often these uses lead to argument and controversy, but at least the LCA model provides scope for factual and reasoned debate and enables both manufacturers and customers to make informed choices between design and product alternatives on the basis of specific effects. Two of the most high-profile recent exchanges have been the war waged over the relative benefits of different forms of milk packaging in Germany, and the dispute in the UK on the environmental effects of disposable nappies (including the challenges by the pressure group Women's Environment Network to claims made for non-chlorine bleached products as being environmentally-friendly).[11]

INDEPENDENT VERIFICATION OF ENVIRONMENTAL AUDITS

The need to verify the validity of companies' accounts through the intervention of outside auditors is widely understood. The discipline of external verification is now being extended to environmental reporting, where external auditors or some other outside authority reviews the audit. This is a trend which began in the 1980s with the work of Arthur D. Little. The value of independent verification was used very successfully in the 1990 Norsk Hydro Report publicity (see 'Public Relations', below), and is now in use by many others, including the Body Shop International in its *Green Book* (see 'Operations', Chapter 8).

11.5 The communication mix

The selection and prioritization of communication channels is key to any public affairs strategy. The development of a holistic understanding of the corporation requires information to be gained across all media. A communication strategy based on stakeholder communication principles seeks to ensure consistency as well as accessibility, legitimacy and external validation. It is therefore critical that, when values, product benefits and opinions on key issues are being promoted, the arguments are internally consistent across all media.

Brand management and communication

The psychological significance and longevity of brands in consumer societies is becoming increasingly realized; where once we might have had other complex social symbols, now an important component of our social signalling system is the brand preference, and in part we create our sense of self through purchasing decisions. For many corporations and organizations, the realization of the power of brands led to a rush for new corporate identities and expensively designed logos in the 1980s. The claim, however fashionable, remains convincing: Western industrialized societies are glued together by their brand icons – from the famous Coke, Schweppes,

McDonalds insignia to the BMW and Range Rover names, the brand is the truly international language of our times. The essential nature of these brands is not simply lettering on a page but the imagery that has so carefully been associated with their product performance. In the hands of a brand marketer, 'caring', 'safe', 'fun', 'young' are no longer simply words but the building blocks of 'the brand personality'. These personalities have now gained such prestige that they may be valued on the owner's balance sheets.

From some environmentalists' perspective, the aura surrounding brands signifies the cynical manipulation of the consumer through advertising hype. This response misses a significant dimension in the consumer cycle: that people purchase to differentiate themselves, to define themselves by associating with the values of these artificially exciting, external personalities. If the sustainability argument is to be assimilated through the developed world, we require a redefinition of the brand personality to incorporate the values and aspirations of environmental good housekeeping. For marketers to ignore this challenge is to risk loss of confidence in the brands they have lovingly nursed along for decades.

Many mainstream packaged goods are rising to this challenge, assimilating environmental concern and responsibility into the overall brand values of their products. This is an alternative strategy to that of developing overtly 'green' products. The movement is in part due to a new sense of corporate responsibility, although cynics might say it has more to do with the threat to market share brought about by quicker-on-their-feet, smaller rivals with an ethical or environmental mission.

Packaging

Packaging performs many functions – to contain, to provide security, to stop spoilage, to assist display, to educate and inform the user. Packaging plays its part in communication in two ways: by what it says and, most important environmentally, by what it is.

Much modern packaging design has failed to acknowledge that there is an overriding need to minimize environmental damage through the responsible choice of materials, printing and adhesive technologies and to ensure appropriate disposal and recycling. Companies with an environmental story to communicate may use packaging as both part of the brand values and as a medium to explain the producers' environmental credentials. Minimization of waste, through the size and concentration of product on offer and material type, offers significant production, handling and disposal cost savings, some of which can create budgets for the further environmental education of consumers. (See example 11.3.)

Procter & Gamble

Since 1989 Proctor & Gamble has been using LCA to improve its environmental performance. Applying LCA to a broad range of detergent and personal care product packaging has resulted in significant reductions in solid waste by reducing packaging by means of new packaging systems, including refills in soft PET pouches and paper cartons instead of standard virgin material plastic bottles, as well as the introduction of bottles with approximately 30 per cent recycled HDPE content. Products have been reformulated to offer triple-strength concentrations. The consumer launches of these changes involved heavy media support and consumer education programmes.

Example 11.3 Proctor & Gamble

Advertising

The highest profile aspect of communication is often the paid-for advertising campaign, usually chosen with extreme care and bearing the highest risk. For many advertisers wishing to exploit the new-found consumer environmental concern, the late 1980s and early 1990s, as we have shown, were not too good. Claims were challenged, knuckles rapped and ads had to be withdrawn. Some agencies even got to the stage of advising clients who had environmental stories to tell not to go ahead. Why? The rush of poor quality communications soured readers' and viewers' attitudes to manufacturers' pronouncements, and the attempt to apply premium-pricing strategies on the basis of environmental credentials was not well-received by consumers, who felt that companies should clean up their act without asking their customers to pay more for the privilege.

PRODUCT PROMOTION

Consumer cynicism poses difficulties for advertisers in the promotion of their selected product messages, and for those with *genuine* environmental benefits there is justifiable concern that money spent will be wasted if no-one is prepared to believe what they have to say. The strategic response, therefore, may be to reschedule the advertising support to a later stage in a communication campaign cycle, after the general issues have been aired more widely, through the use of public relations and awareness campaigning, rather than talk about product benefits that are not currently understood by the consumer. Adopting this route forward also has the benefit of avoiding castigation by the regulators.

For those with newly-found environmental claims for existing products, it would be wise to address the whole issue of credibility and legitimacy to ensure that the consumer understands that this product change is the result of substantial alterations in the way the company does business, not simply a 'green wash' or a one-off product in a largely unchanged range.

CORPORATE ADVERTISING

The use of environmental performance in corporate advertising is a phenomenon that is set to grow as marketing departments' interest returns to the area and as companies are forced to compete in a business-to-business setting on the basis of their environmental standards. The increased sophistication of both consumer and business audiences in the 1990s will be a challenge to communication professionals to ensure that a gearing-up takes place, replacing the hackneyed 'trees/birds/fishes/blue sky/meadows/forests' approach and focusing instead on tangible progress and provable achievement. In this, corporate advertising will be forced to keep pace with the anticipated changes in corporate reporting and environmental profiling.

Sales promotion and sponsorship

The whole sales promotion arena is a difficult one to justify on environmental grounds. Beyond the justification of brand-switching there can be few commercial activities that are so blatantly aimed at more and more consumption! Both the rationale and the techniques used, from the direct mailshot to mountains of consumer literature, do not accord well with corporate commitments to reduce waste and to use fewer material resources. Short-term commercial considerations may necessitate the continuation of these techniques, and mitigation measures may be all that are possible in the short term, for example from the use of recycled paper or the creation of reusable in-store materials. In the long term, however, any business which is dependent on the manipulation of consumer spending habits through expenditure of this nature cannot be said to be truly sustainable, and the relationships both with customers and other stakeholders in the business should be reconsidered.

Public Relations

The field of public relations has often been seen as slightly shady by outsiders, with the lobbying and media relations work of this aspect of communications as very suspect. Yet for a corporation bent on real change towards a more environmentally friendly way of working, promoting products with significant benefits and which aim to educate the consumer, there can be no better use of communication expenditure. The profile-raising implications of environmental activities as a means of standing out from the crowd should not be ignored after the notable success of those who have adopted a leadership stance. (Example 11.4).

Norsk Hydro's experience

Norsk Hydro is a Norwegian-owned international chemicals-to-energy conglomerate. The mid-1980s was a difficult time for the company, due to a number of environmental incidents and the active campaigning of environmental groups who discovered mercury soil contamination around the Heroya chlor-alkali and the Porsgunn chlorine works in Norway. This public exposure affected Norsk Hydro significantly, being one of the largest employers in Norway, a country with a high level of interest in environmental matters. The company chose to respond positively to these experiences.

High-level commitment was given to annual audits of all companies and plants within the group, and eventually this led to a published report. The initial Norwegian report was circulated to Hydro employees, local and national authorities and homeowners close to plant sites, also to schools and colleges. In 1990 a more comprehensive report covering worldwide operations was released. Hyrdo believes that this public communication was responsible for the rapid manner in which it was able to regain public confidence after this period of difficulty, culminating in the receipt of the Norwegian award of 'the best company to work for' in 1990.

The UK operation, despite leadership and a significant share of the relevant markets, 'was almost invisible to the public eye'. Environmental reporting was seen as a way of both communicating about an important issue and of raising the public profile of the company. What made the UK initiative so innovative was that management decided it would publish the contents of the audit whatever the findings, and that it would commission an independent third party to verify the audit results. The UK report was published in October 1990 and over 10 000 copies of the report have been circulated, including recipients in 24 countries. The company's profile has been raised significantly by this activity[12].

Example 11.4 Norsk Hydro

Public relations activity is a complex area, one which itself splits into a wide range of functions – media relations, consumer education/product placement, trade/peer group relations, sponsorship and charity work, shareholder and financial PR. Each of these may require a new approach by the environmentally focused corporation.

The growth of new consumer and business-to-business environment magazines, and the need for environmental information to be presented to existing publications, necessitates a fresh approach to media work. Consumers have prejudices about environmental questions that often defy rational explanation; these can be challenged through awareness campaigns and product placements. The trade press may be just as slow to realize the relevance of environmental improvement, and taking a leadership role in an area has strong potential payback for the company or companies that take up the challenge. Similarly, the need to communicate with shareholders and other stakeholders through financial publications and documents gives ample opportunity to detail the company's environmental achievements. Profile-raising activity in this area has a major role to play in the business's overall strategic response to the environment and meeting the financial demands which this places upon the company, by encouraging investor support.

CHARITABLE GIVING

For companies with sizeable budgets for charitable expenditure, or for those who are often approached because of their profile in this area, processes are required for managing the selection of suitable candidates. The environmental aspirations of the company and the need to support and reward stakeholders require evaluation of the gift criteria. Given that demand will always outstrip supply of charitable funds, it is important to ensure that you have adequate information and an agreed process for evaluating proposals in a manner which meets the overall objectives of the organization.

CRISIS COMMUNICATIONS

Nowhere is integrity in communication needed more than in the field of crisis management. Ensuring that errors are dealt with effectively, that statements made show respect for the feelings of those who have been put at risk, and that damage is mitigated as much and as soon as possible, is of vital importance. Many of the highest profile environmental accidents have their counterparts in terms of damage, but the notoriety associated with the incident has often as much to do with the mishandling of information and a lack of corporate backbone as with the extent of the damage sustained. The press response to Exxon at the time of the Exxon Valdez disaster is a prime example. Time and place undoubtedly play their part, but any organization with pretensions to environmental responsibility operating in a high-risk situation will ensure that it has a fully operational code of practice for dealing with the communication implications of a critical error. The number of sectors that this applies to is growing all the time, and complacency is a dangerous fault in this area.

ENVIRONMENTAL AWARDS

There are a growing number of environmental awards covering reporting, practice and innovation. Both participation in the awards and sponsoring give opportunities to increase the environmental profile of the organization. Conversely, being awarded a 'Green Con' type award for failure in one product category or site can distort the overall perception of a company's environmental credentials, which in some cases may be an unfair reflection on overall efforts.

11.6 Corporate space

There are opportunities to promote environmental achievement or the company's strategic alliances using the company's sites themselves and delivery vehicles. This 'free' medium is used to good effect by the Body Shop International, which has made much of the advertising sites that are its own shop windows and the sides of delivery contractors' vehicles. Using in-house display space to promote and provide a focus for corporate achievement is also important. Some organizations have gone as far as creating wildlife gardens and wild flower areas in their grounds as a staff resource.

11.7 Corporate environmental profiling

The need to account publicly for the environmental performance of the organization has spawned a growing number of corporate environmental profiling options, as we detailed in Chapter 5. For some companies, because of the size and complexity of their operations, or the level of their aspirations and profile in this area, *all* would be appropriate. For the communication or public affairs director, knowing which route to take at what stage of the company's evolution in environmental response is a difficult judgement, given the resource and financial implications. The complexity of much of the information involved will require close collaboration between communication, environmental management and financial personnel to ensure the best possible outcome for the organization.

11.8 Alternative communication strategies

We have outlined the range of communication issues raised by the development of a strategic approach to the environment, and it will therefore come as no surprise to learn that different companies have responded to the challenge in a wide range of ways. Their response is occasioned as much by their sector, level of awareness and culture as by the external progress being made elsewhere.

Defensive/damage limitation

Where the environment is seen as a challenge to performance, profitability or simply 'business as usual' the first line response is defensive, responding to interest and questioning as a threat. Supporters of this strategy are responding to external pressures by challenging environmentalists' claims,[13] advocating market sovereignty and lobbying for deregulation on environmental matters. Internally, too, some companies have instigated channelling procedures to ensure that environmental matters are handled by a single individual, not to improve the availability of information but to act as a block or filter. Although there may be an absence of environmental enthusiasm, a bland and commonplace statement about environmental principle may be found in the company report. At best, such responses, if conscious, are the result of ignorance, inertia and prejudice, or these organizations have something significant to hide. Alternatively, they may be well aware that the environment cannot be handled within their current organizational framework, and their response is fuelled by an understanding of how unsustainable their business operations and culture really are.

Reactive/consumer-focused

Others are reacting to the environmental agenda only as far as the consumer, public pressure or legal requirements force a response. The availability of a general policy statement, the presence of a supporting introductory guide, but no formal targets for improvement are the hallmarks of this strategy. Here, consumer and public pressure needs to come right into the corporate doorway before there is a strategic response, and by the time attention is focused on the organization, strategic choice gives way to panic-ridden pronouncements, which jar with both the culture and the business strategy of the organization. The inclusion of non-quantified environmental references in the statutory reports is to be anticipated.

Environmental competitive advantage

Here, the decision to make significant environmental changes is an answer to the commercial opportunities afforded, to take advantage of the benefits of enhanced corporate reputation over one's competitors or of the product or brand advantages to be had by provable environmental advantages. The emphasis will be on product reformulation or innovation and strong clear communication statements, covering product changes, ecolabelling and corporate activities, through public relations and advertising. An investment in the work needed to generate a comprehensive environmental report, and the subsequent improvement targets, would be consistent with this strategy.

Environment as profile raiser

The company which is determined to use environmental performance as leverage, raising its profile with target audiences from consumers to financial analysts, has made the decision to be on par with the current best practice generally or for its sector. This route has been used successfully by piloting the environmental reporting techniques outlined, and through participation in the growing number of 'round tables' and special interest groups in support of industry's response to the environment.

At worst, where this strategy has proceeded from calculation divorced from any genuine commitment, there has been an emphasis on highly visible activities, from corporate advertising campaigns through to glossy but essentially vacuous public statements which may later be exposed through pressure group activity.

At best, the visibility sought has led to real reappraisal of the issues involved, as communicators struggled to find the necessary supporting information, raising the profile of the environmental issue in the company and in turn causing changes to operating practices. Many of the companies which used profile-building in the late 1980s have now matured into a realization that the environment cannot be 'milked' in this way, and with that awareness has come a deep-seated change of internal priorities.

Environment as a communication platform

This is based on the business strategy of environment as a strategic opportunity, with the company actually setting out to solve environmental problems for industry, society or consumers. A company working to this type of agenda would need a fresh communication style, one which talks to people about their needs in a different way than the conventional techniques used to generate wants. It is likely that such an approach would be highly targeted, focusing on niche markets to identify and respond to particular circumstances. The overall emphasis is on working in a two-way partnership.

Promoting sustainability

For the committed organization which has fully accepted the need for a complete refocus on achieving sustainability, the communication strategy will cover two dimensions: first, an outward focus on educating, informing and joining with consumers and stakeholders in campaigning for change, and secondly, an internal effort to motivate, empower and transform the organization to make the necessary changes. This strategy can often be confused with the self-promotional element of 'environment as a profile raiser', but it can be differentiated by the emphasis on total honesty, rigorous challenges to continuous improvement and the willingness to give the means to make changes to others. An example would be the support given by the Winter organization for the promotion of their Integrated System of Environmental Business Management through the *Bundesdeutscher Arbeitskreis für Umweltbewußtes Management*, BAUM (German Environmentalist Management Society). Here the emphasis is on giving away information, as well as the means by which others can 'empower' themselves, often for little formal thanks or praise.

References and further reading

1. Richard Adams, Jane Carruthers and Sean Hamil, *Changing Corporate Values, New Consumer*, Kogan Page, 1991.
2. Stephen R. Covey, *The Seven Habits of Highly Effective People*, Simon & Schuster, 1990.
3. See John R. Beaumont, Lene M. Pederson and B. D. Whitaker, *Managing the Environment*, Butterworth-Heinemann, 1993, page 130 for examples.
4. Neil Verlander, '*Pressure Group Perspective*', in M. Charter (ed), *Greener Marketing*, Greenleaf Publishing, Sheffield, 1992.
5. The Code and Guidelines Relating to Environmental Claims in Advertising, Incorporated Society of British Advertisers, 1991.
6. MEDIA Natura Bulletin 10, April 1992.
7. ENDS Report 199, pages 15–18.
8. ENDS Report 230, page 230.
9. Waldemar Hopfenbeck, T*he Green Management Revolution*, Prentice-Hall, 1992 (Germany) 1993 (UK).
10. Walter Coddington, *Environmental Marketing*, McGraw-Hill, 1993.
11. 'No end in sight for milk packaging war', *Environment Business*, 20 October 1993, page 1/ Warmer Bulletins Nos. 36 and 37, Warmer Campaign, Tunbridge Wells, Kent, UK.
12. Adapted from S. Schmidheiny (ed.), *Changing Course*, pages 197–201.
13. For example by promotion of the views of Richard Balling, Director of the Office of Climatology, State of Arizona, who rejects the global warming hypothesis.

PART FOUR
Implementing environmental strategy

In order to implement both the corporate and the departmental environmental strategies in an integrated way it is necessary to bring them together into a total picture, both conceptually and organizationally. The two chapters in this section are designed to do this.

Chapter 12 sets out the requirements for an environmental organization that can turn the strategic choices into realities.

Chapter 13 puts together the pieces of the jigsaw of corporate and functional choices into a coherent picture of the emerging sustainable business.

12 Organizing for environmental change

1. How a company organizes itself to put its environmental strategy into practice is a strategic choice in its own right, and the success of the strategy depends on it.
2. This requires answers to a number of specific questions.
3. The first of these is to decide the mission of the environmental organization, or what difference it is intended to make to the company. There is always a choice of quite diverging missions, and different missions require different forms of organization.
4. Next, responsibility for a range of functional tasks required by the strategy and mission has to be allocated to departments and individuals.
5. Managerial accountability for planning, allocating and reviewing all this work has to be similarly apportioned.
6. An important aspect of this process, which is often overlooked, is deciding on the appropriate organizational levels at which accountability for the various tasks and objectives should be pitched. We provide a model on which these decisions can be based and discuss in detail the levels at which different roles in an environmental organization can most effectively be performed.
7. Should environmental strategy be implemented 'from the top down' for control and consistency, or 'from the bottom up' for ownership and commitment? A combination of the two may gain the best of both worlds.
8. The organization must also ensure that all environmental activities are properly coordinated and integrated across the company, and it must decide and define the pivotal role of the environmental manager.

12.1 Organizing for environmental management is a strategic issue

The success of any strategy depends on there being an organization to deliver the goods. Organization structures are geared to the objectives to be achieved, and when new objectives are set by a change in strategy an organization has to be created to achieve them.

To be effective, an organization structure has to perform two tasks, summed up in the twin slogans:

Divide and Rule
Connect and Energize

The criteria of successful environmental organization include not only the ability to maintain control over environmental effects by establishing accountable roles, monitoring systems and review procedures, but also the capacity to integrate and empower the efforts of different parts of the enterprise towards total company-wide solutions and goals.

For those companies who take the strategic view of the environmental scenario that we advocate in this book, the way they organize both for ongoing environmental compliance and improvement, as well as competitiveness, opportunity and ultimate sustainability, is a strategic question in its own right.

12.2 Questions to be answered

There are a number of quite specific questions that have to be considered if the organizational arrangements for environmental management are to be more than haphazard:

- How is the mission of environmental management defined? How innovative should it be?
- What functional tasks arise from this mission? (i.e. work related directly to achieving the required environmental performance)
- What managerial tasks have to be performed? (i.e. setting objectives and ensuring they are met)
- Where should accountability and authority for these activities be located, both in terms of new or existing roles or functions and of organizational levels?
- How should environmental strategy be coordinated and integrated across functions, subsidiaries or business units and sites?

The wide range of organizational arrangements to be found in existing companies represent different ways of answering (or failing to answer) questions like these. The remainder of this chapter is devoted to looking at these answers.

12.3 Defining the environmental mission

A prerequisite for successful organization design is to identify the 'mission' of the organization in question. This is defined by Organization Development (OD) specialists as the difference the organization is designed to make by processing a given set of inputs and turning them into outputs. For example, the mission of a hospital might be defined as turning sick people into well people.

Except that in most cases there is a choice of missions – a teaching hospital might define its primary mission as turning medical students into competent doctors. Or the two might be combined into an overarching mission of optimizing the standard of healthcare for the community. Each of these mission definitions will have different implications for decisions on how to organize.

In designing the environmental organization of a company, the same question has to be asked: What difference do we want it to make to the enterprise as a whole? The environmental mission is closely related to the strategy. The following examples of different environmental 'missions' are easy enough to relate to some generic strategies discussed in Chapter 4.

A. The mission is to keep the regulators off our back.
B. The mission is to ensure that we comply with the legislation.
C. The mission is to minimize, or at least limit, our adverse impacts, within reason.
D. The mission is to develop our business by being greener than our competitors and/or solving environmental problems for other people.
E. The mission is to ensure that our company makes a significant positive contribution to global/local environmental problems.
F. The mission is to become a sustainable company.

Decisions about environmental strategy or mission are likely to affect the most appropriate way to organize for environmental management. For example, the definition in B implies that responsibility for identifying priority issues should be given to a specialist middle manager, while E would require this responsibility to be carried by a senior manager, possibly the CEO, with professional advice, and D calls for the creation of a cross-unit senior management group making recommendations to the board.

Adopting mission D, E or F would involve a degree of innovation which could not easily be handled by a single role, however senior, but requires creative interaction between a number of people in settings (such as project teams) which facilitate creativity.

12.4 Allocating functional tasks

When companies decide to take environmental protection on board, they find themselves facing a whole series of potential new tasks. A central element in creating the environmental organization is deciding which of these tasks to take on and what roles or functions to make responsible for them.

Meeting statutory requirements

Discovering how much of the growing and proliferating body of environmental legislation applies to them and what they need do about it is the most pressing need at the outset for companies in environmentally sensitive industries. This helps to explain why in so many companies environmental responsibility has been added to the existing Health and Safety manager, who was already concerned with legal compliance in relation to safety and UK Control of Substances Hazardous to Health (COSHH) regulations.[1]

Tasks under this heading include:

- producing inventories of applicable legislation and toxic substances
- drawing up applications for permits and consents
- waste documentation under Duty of Care.

But companies who take a strategic approach to the environment find that legislative compliance is only a part of the total picture. Here are some of the others:

Auditing and monitoring, including:

- environmental auditing of sites
- drawing up environmental effects inventory
- performance measurement and reporting
- environmental assessment of suppliers and contractors.

Pollution prevention and waste minimization
Development work, including:

- process development (clean technology)
- product development/stewardship.

Environmental training
External relations, including:

- public relations/corporate affairs
- representation on industry and public bodies.

Looking ahead; anticipating environmental pressures

How many and which of these tasks the organization will take on will depend on its environmental strategy and mission, explicit or 'default'. In so far as these tasks are taken on board, there has to be serious thought about where the responsibility for them should lie. In particular, the same logic which made a health and safety function an appropriate home for the legislation-related tasks may well suggest the need for a different location for some of the more developmental and strategic tasks in the list above.

12.5 Allocating managerial tasks

In an organization of any size, a strategic approach to the environment involves more than deciding on the concrete jobs to be done and who is to do them. It is also important to put in place an adequate management superstructure to give direction, continuity and control to the actions to be taken. The managerial tasks that have to be taken care of include the following.

Strategic choice

- formulating strategy
- identifying priority issues and requirements.

Implementation

- translating strategy and priority issues into policy, objectives and targets
- allocating resources to objectives
- setting up management systems.

Execution

- planning to achieve objectives and targets
- reviewing and monitoring.

All of these are, of course, ongoing tasks, so the responsibility for each is not a one-off, but includes regular review and updating.

12.6 Locating accountability

Locating who should carry the accountability both for the functional and the managerial elements of environmental management is a two-dimensional task, of selecting not only the most appropriate function or role for each, whether in line management, a specialist area or a separate environmental function, but also determining the appropriate *level* in the hierarchy.

Levels of responsibility

The levels of managerial responsibility have been found to be one of the most important matters to be determined in designing structures for maximum effectiveness. In most organizations it is possible to identify five distinct levels of responsibility and decision making which are essential to the effective functioning of the enterprise.[2] In very simple terms these can be defined as shown in Table 12.1.

Table 12.1 Levels of responsibility[3]

Level	*Responsibility*	*Example*
L5	Defining mission and strategy	Chief executive
L4	Deciding and deploying required resources and plans	Director
L3	Setting up systems, rules, guidelines for producing results or outputs	Department manager
L2	Applying systems, rules, guidelines to specific situations to decide results required	Supervisor, technical specialist
L1	Producing required results	Operator

The decisions to be made at each of these levels are not just different in 'size' or scope but different in kind, and require different outlooks and capability. In an area where one of these levels is missing, certain decisions either do not get taken at all, or are not given the kind of thought they require. This is just as true when responsibility is at too high a level as when it is too low.

For example, if environmental strategy is made at Level 3, a typical level for an HSE or environmental manager to be appointed, the chances of fully integrating it with the business strategy of the company will be reduced, and the strategy may be limited to a set of guidelines for dealing with ongoing situations, rather than representing a considered decision on where the company wants to position itself.

Conversely, the board, or the director given responsibility for environmental issues, may decide that the job of translating the company-wide environmental policy into practice should be left to individual line managers. If, as a result, no Level 3 manager is given overall responsibility for identifying key environmental risks and setting up systems and guidelines for dealing with them, there is a high likelihood that important requirements will be overlooked and the company's environmental performance will be at best inconsistent and unreliable.

Environmental roles

The five-level model in Table 12.1 can be used as a convenient framework for discussing the main environmental roles which companies should consider in relation to their overall environmental strategy. Whether or not all these roles are needed by every company is open to debate, but in omitting particular roles it is useful to consider the decisions and actions which might go by default in their absence.

LEVEL 5: DECIDING ENVIRONMENTAL STRATEGY

When Edgar Woolard, as CEO of Du Pont, described himself as his company's Chief Environmental Officer, he was making an important point. The environmental strategy and mission of a company has to be decided by the chief executive in conjunction with the board, otherwise it goes by default or becomes subsumed under another function's strategy. Thus environmental strategy can be made the inadvertent hostage of financial strategy, by omitting to consider whether the same payback requirements should apply to environmental as to other investments. Or it is possible to make environmental strategy an adjunct of marketing strategy, by defining it as a response to customer requirements.

For example the corporate planning manager of a specialty steel company found that only two of its 300 customers had asked any questions about the company's environmental record. He concluded that the environment was not an important strategic issue for the company.

Failure to address the environment at a strategic level has been evident in some of the country's largest financial institutions, who have confined their interest in environmental issues to low-level operational questions such as recycled paper, apparently ignoring the potential effects of their investment policies. The BCSD gives an example of the powerful impact such policies can have:

> For example, the net value method discounts projected investment returns 20–50 years from now so heavily that they are virtually valueless. This is especially true in developing countries, where the uncertainty of the investment climate pushes the discount rate up. This often results in rapid consumption of resources rather than investment aimed at increasing the overall resource/result ratio. Scarce capital can be attracted to enterprises or projects that undermine rather than promote sustainable development.[4]

In the absence of Level 5 involvement, there are two possible ways in which strategy is decided. One is for the functional or unit heads at Level 4 to develop their own functional strategies, on the basis of some of the considerations we have outlined in the previous chapters.

The other possibility is not to address the strategic question at all, by giving responsibility for the corporate environmental response to a director or manager at Level 4 or 3. Such a manager, if very enterprising and far-sighted, might take on the job of drafting strategy him or herself and push a set of recommendations back to the board for endorsement. The more likely outcome, however, is the adoption of the 'default' strategy of compliance with legislation, plus support for any functional strategies that happen to be around.

LEVEL 4: PLANNING AND RESOURCE ALLOCATION

Any company which attaches importance to the proper implementation of its environmental strategy needs to nominate a senior manager – a director or equivalent – who is responsible for putting in place the necessary plans and resources which alone can turn the strategy into reality.

Often this role is carried by an existing director – technical, perhaps (because of the technological implications), or personnel (because of the health and safety connection), perhaps marketing (if competitive considerations are uppermost), or purchasing (where, as in some retail groups, the environmental spotlight is on suppliers). In such cases the director normally requires specialist or Level 3 support to work out and monitor ways and means of implementing the strategy beyond the boundaries of his or her function.

In companies where a Level 4 environmental role is missing, where, for example, a 'green' CEO delegates directly to an Environmental Manager at Level 3 or even a specialist who functions in typical Level 2, reactive, trouble-shooting mode, it becomes very difficult to institute a coherent and effective environmental improvement programme.

This is not necessarily a question of competence. Even where the environmental manager is able to design an excellent plan, 'political' pressures, both internal and external, are likely to throw it off course. Internally, the environmental manager comes up against conflicting interests of both Level 3 and Level 4 managers, while externally he or she is taken up with fielding regulators' and other demands on the CEO's behalf.

LEVEL 3: CREATING THE ENVIRONMENTAL MANAGEMENT SYSTEM

These days every company of any size requires a formally worked out system for managing its environmental impacts, whether or not it is part of one of the more obviously 'polluting' industries. Table 12.2 illustrates the

Table 12.2 Industries in the firing line 1960s–1990s[5]

1960s	*1970s*	*1980s*	*1990s*
Coal (mining and air pollution)	Aerosols	Aerosols	Aerosols
Detergents	Airports	Agriculture	Agriculture
Mining and quarrying	Asbestos	Airports	Air conditioning
Pesticides (e.g. DDT)	Automotive fuel efficiency	Automobile exhaust emissions (e.g. lead)	Airlines and airports
Water (dams)	Biotechnology (accidental release)	Biotechnology (deliberate release)	Animal testing
	Chemicals (e.g. dioxins, PBBs, PCBs)	Coal	Armaments
	Coal (mining and air pollution)	Computers	Automobiles (fuels, cars)
	Deep sea fishing	Deep sea fishing	Banking
	Detergents	Detergents	Biotechnology
	Heavy lorries	Fertilizers	Catering
	Metals	Forestry	Chemicals
	Motorways	Incineration	Coal
	Nuclear power	Insurance	Computers
	Oil tankers	Landfill	Crematoria
	Packaging (e.g. glass bottles)	Motorways	Deep sea fishing
	Passenger jets	Nuclear power	Detergents
	Pesticides/herbicides	Oil tankers	Dry cleaning
	Pulp mills	Onshore oil and gas	Electricity supply
	Tobacco	Packaging	Electrical equipment
	Toxic waste	Paints	Fashion
	Transport	Pesticides/herbicides	Fertilizers
	Water	Plastics	Fish farming
	Whaling	Pulp and paper	Fishing
		Refrigeration	Forestry
		Supermarkets	Incineration
		Tobacco	Insurance
		Toxic waste	Investment
		Tropical hardwoods	Landfill
		Tuna fishing	Meat industry
		Water	Mining
		Whaling	Motorways
			Nuclear power
			Office supplies
			Oil tankers
			Onshore oil and gas
			Packaging
			Paints
			Paper
			Pesticides/herbicides
			Plastics
			Property
			Pulp and paper
			Refrigeration
			Schools
			Shipping
			Supermarkets
			Textiles
			Tobacco
			Tourism
			Toxic waste
			Transport
			Tropical hardwoods
			Tyres
			Water

extent to which the focus of environmental attention has broadened over the past 30 years to take in more and more industries to a point where very few can any longer afford to remain aloof.

The concept (and name) of an environmental management system (EMS) has been popularized by the British Standard BS 7750, but is independent of it. Whether or not a company intends to apply for certification under BS 7750, it cannot ignore the purposes for which an EMS is set up. In order to ensure that the effects of its operations on the natural environment are known, monitored and controlled at a level which is acceptable to both its social and ecological 'stakeholders', the company needs a set of documentation, procedures, measurements, controls and review processes which is tailored to its particular processes and circumstances.

Designing and setting up such a system has to be undertaken by someone who is used to operating at the interface between planning and execution, between theoretical understanding and practical outcomes, which is typical of Level 3. In the absence of such a role, a system which is established from a Level 4 or Level 2 position runs the risk of being either unworkable or incomplete.

The environmental manager responsible for a single large site or a number of smaller ones is the central Level 3 role who has to create the EMS and without whom it risks being ineffective. But as the point of the EMS is to control the environmental impacts of the line and technical operations of the company, the role of the managers responsible for those operations is equally vital. Unless they build the EMS into their own ongoing management practice, it cannot achieve the objectives for which it is created. What this means is that environmental protection has to be formally built into the roles of these managers, with the official EMS supporting them in meeting standards and targets which form an inherent part of their own jobs.

LEVEL 2: DELIVERING ENVIRONMENTAL IMPROVEMENTS

According to the precise strategies and plans established at Levels 5 and 4, there is likely to be a call for roles to be established at Level 2 to carry out specific tasks in such areas as process or product design, supplier assessment, pollution monitoring and reporting. A particular area where important contributions have been made by problem-solving Level 2 appointments is that of waste minimization.

> A company manufacturing consumer products has appointed an engineer working for the Plant Services Manager as Waste Controller for the site. He is responsible for waste disposal (including Duty of Care) and waste minimization by reduction, recovery and recycling, and for maintaining ongoing records of all forms of waste by type, source and destination. His

job is to develop all possible means for reducing the waste, both toxic and non-toxic, liquid or solid, leaving the site for transport to landfill or incineration or as effluent, against corporate targets. The Waste Controller's activities have included introducing a system of bins for different types of waste, removing and recovering reusable chemicals from aqueous effluents and finding outlets for recycling many of the waste materials produced, and since his appointment there has been a substantial reduction in the waste sent for disposal.

Centralization or delegation?

One of the questions which an environmental strategy has to address is how to energize implementation. Given an environmental policy and a set of principles, should implementation be driven down through the organization from the corporate centre ('top down') or should the practical application of the policy be worked out from the grassroots in operating sites and departments ('bottom up')?

In favour of top-down implementation is the ability of corporate management to set tough targets and support them with financial resources where required. Inconsistency in application between different sites or business units could weaken the overall impact of the policy and compromise expected competitive gains. A strong impetus from corporate management, and in particular a willingness to commit resources, sends an unmistakable message to the cynics and sceptics to be found in every company.

Conversely, if subsidiaries, business units, sites or departments are left to work out their own applications of corporate policy, their commitment and 'ownership' will be that much greater. At the same time, they will be in a position to concentrate their efforts in the areas of greatest impact and make beneficial trade-offs between competing priorities.

A US-owned multinational manufacturing and marketing a well-known range of household products in over 40 countries worldwide implements its ambitious environmental policy and principles by a combination of 'top down' and 'bottom up' methods. The Executive Committee, through the Corporate Environmental Council and the corporate environment director, sets specific objectives for subsidiaries, with target dates for their achievement. These are partly in the form of absolute levels to be achieved (e.g. percentage of waste recycled), partly improvements against a reference period (e.g. percentage reduction in source material), partly qualitative (e.g. phasing-out of suspect materials) and require regular performance reports.

In parallel with the corporate target-setting exercise, and sometimes ahead of it, individual subsidiaries with a particularly strong environmental commitment carry out environmental reviews and audits of their local practices and develop their own environmental plans and initiatives, as well as exchanging information with other subsidiaries and the centre. Their efforts are directed at meeting corporate targets where they exist, but are not necessarily confined to the areas for which targets have been set.[6]

12.7 Integrating environmental strategy

Once a company goes beyond a pure compliance strategy, the coordination and integration of environmental strategies of business units and functions becomes an important issue. This has been stressed throughout this book, as we have examined possible strategies for different business functions.

Coordination can be carried out through an individual appointment, such as a corporate environmental (or environmental strategy) director, manager or coordinator, or through a council, committee or steering group on which the different functions and/or business units are represented.

In most of the examples we have come across, both a coordinator and a coordinating group exist side by side, with the individual role-holder acting as secretary, resource person or chairperson to the group. The group sometimes has a purely information-sharing role, but more often it is charged with developing action plans and new initiatives within existing strategies, and often with making recommendations to the board.

12.8 The role of the Environmental Manager

Of all the roles and groupings we have mentioned, the Environmental Manager at present holds the centre of the stage. The recent survey by ENDS[7] shows environmental managers as being fairly evenly distributed between site, divisional and corporate locations, and involved in a wide range of work, from basic statutory compliance to strategic development of new functional environmental policies for product development and life-cycle analysis or supplier policies and continuous improvement.

There are indications that the statutory work is diminishing as more of it is directly taken on by operational line management, and that there is an increasing tendency for more senior environmental managers in larger firms to take on a more strategic role. But it seems that for the foreseeable future the central role of environmental managers will be in setting up and developing increasingly comprehensive environmental management systems for their organizations.

References and further reading

1. According to a recent ENDS survey of environmental managers, almost 50 per cent also have health and safety responsibilities. ENDS, *Environmental managers in business*, November 1993.
2. The organizational model used in this section comes from the work of Rowbottom and Billis and the Brunel Institute of Organization and Social Studies (BIOSS). Their analysis has proved helpful in unravelling management problems in many different kinds of organization. In large organizations, such as groups of companies or business units, with separate missions and strategies, another two organizational levels may have to be added. See R. Rowbottom and D. Billis, *Organisational Design: The Work Levels Approach,* Gower Publishing 1987.
3. Based on R. Rowbottom and D. Billis, op. cit.
4. Schmidheiny/BCSD, *Changing Course*, 1992, page 61.
5. J. Elkington, P Knight and J. Hailes, *The Green Business Guide*, Gollancz, 1991, page 15.
6. Condensed from K. Knight, 'Proactive and Responsible: One Company's Path from Green Policy to Green Practice', *Business Strategy and the Environment*, 1, 4, 1992.
7. ENDS, op. cit.

13
Building a sustainable business

1. This chapter tries to map out the different aspects of a sustainable business strategy.
2. The goal of such a strategy is a business which is both environmentally and economically sustainable in the long term.
3. To move towards this goal will require a new conception of management responsibility to a greatly extended system of multiple stakeholders with multiple performance criteria.
4. The choice of a coherent environmental strategy is critical, as is its implementation through an effective strategic planning process.
5. New accounting systems, which track both the costs of environmental action and the physical flows into and out of the business, will be needed to provide the basis of sustainable management.
6. The environment is likely to occupy a central place in a coherent corporate marketing strategy.
7. R and D will be called on to spearhead new responses to environmental needs and to the emerging business opportunities of sustainable development.
8. Operations management is in the front line of the drive for environmental sustainability, and its ability to deliver the necessary changes depends on a sound operations strategy.
9. Procurement and logistics strategies will play an important part in minimizing environmental impacts over the whole product life cycle.
10. The achievement of these goals will depend on the capability of *people* to bring about the necessary changes. This will call for a

much greater involvement of Human Resource Management than hitherto.

11. Emerging interdependence of the business with a greatly enlarged stakeholder set needs to be reflected in a two-way, multi-channel, consistent and validated communication system.
12. The ability to implement effective environmental and sustainable management depends on putting in place a structure of accountable roles at the right organizational levels.

(Note: The numbering of sections in this chapter corresponds to the numbers of the chapters dealing with the specific topics.)

13.1 Putting the pieces together

At this point it is time to draw the threads together into a coherent picture of the sustainable business and the processes for bringing it about. We said at the outset that we did not intend to prescribe but to highlight choices. Whether we have always succeeded is for the reader to judge. We certainly have not tried at any point to conceal our own bias in favour of the long-term sustainable option – how could we, when we see it as the only escape route out of our present predicament that would not *necessarily* make matters even worse than they already are for both the developed and the developing world.

But at the same time we have tried to hold back from presenting 'all or nothing' choices, knowing that circumstances alter cases and that there may be more than one way to reach a common goal. The task now, however, is a different one. Although every organization is unique and must find its own path to sustainability, there is a need to recall the important components of the sustainable business as they have come to light in the course of our exploration of different aspects of the typical enterprise. In the following attempt to assemble the pieces into a meaningful whole, the result may not be the only possible one, but at least we hope it is consistent, coherent and makes sense.

13.2 The goal

First of all then, the goal of our endeavours, the *sustainable business*, is one which, as in the past, satisfies society's needs and wants, but realizes that these now include, as a matter of priority, the need for 'sustainable development', meaning economic development towards satisfying the world's unmet needs within a clean and stable environment. In practice, it:

- satisfies existing needs while using resources sustainably

- maintains an equilibrium in its relations with the natural environment, based on clean technologies, continuous reuse, recycling or renewal of resources and restoring any damage caused
- contributes to solving the problems of system overload which human expansion has created, rather than exacerbating them
- generates enough income to sustain itself.

13.3 New responsibilities

The argument of this book has been that companies will not achieve these goals simply by wishing to do so, but that survival and successful adaptation will demand a considerable programme of strategic development of all areas of the business in order to gear them to the requirements of sustainable operation.

As the basis of such development, we have suggested that there is a need to move to a new conception of management responsibility, involving:

- an enlarged set of 'stakeholders', meaning all those who may be affected by management's decisions, whether supermarket shoppers, New York stockbrokers or Pacific tuna fish
- an extended model of the 'system' being managed, which would include not only the people on the payroll, your customers and competitors and the tax man, but such other interested parties as rainforests, endangered species and the ozone layer
- multiple performance criteria to represent the complex system of flows of money and materials, contributions and benefits, air, water and contamination, into and out of the organization
- a collaborative harnessing of creative energy from inside and outside the organization, on a basis of mutual trust.

It may be argued that all this is idealistic pie-in-the-sky which takes little account of the realities of stock markets and the legal and financial frameworks which dominate business. Whether that objection is valid or not depends not on us but on the perceptions of the leaders and power-brokers of the world's largest companies. If they are sincere in their espousal of sustainable development as an overriding objective, then they have the power to shift any institutional obstacles that stand in the way. If not ...

13.4 Choosing and planning the strategy

To survive in the business environment it is necessary to get your business strategy right. We have argued that the company's approach to the natural environment, its *environmental strategy*, needs nowadays to be an integral part of this. We have examined a number of positions which companies can take, and some of them are not, of necessity, mutually exclusive.

Thus it is possible to move towards long-term *sustainable operation* either through a position of good *corporate citizenship* or by seeking out the *strategic opportunity* linked to environmental protection, or even by a combination of both. In our own view, a strategy which treats environmental pressures as more of an opportunity than a threat is likely to be both the most profitable *and* the most sustainable, because it avoids from the outset the sense of inherent conflict between commercial and environmental requirements that is the source of many of the present *defensive* strategies.

The decision to accept sustainable development, however, as a goal provides no more than a point of departure for planning, and certainly not a route map. Taken together with the necessity to integrate environmental sustainability with a viable *business* strategy for the organization, it presents strategic planners with a considerable challenge.

In the conditions of growing turbulence affecting business in the present period, of which developments in environmental concern represent a significant aspect, we suggest that planners would benefit from considering a wider range of methods than they may have employed in the past, both to model the diversity of business contexts against which their strategies might have to be tested and to achieve a better balancing of a much larger set of stakeholder needs and inputs than they may previously have had to consider.

13.5 Accounting for the environment

It is important for companies to realize, far more than most have done up to now, the key role their accountants will have to play in the move towards sustainability. In the coming decade or so, business is going to have to face up to the progressive withdrawal of the 'environmental subsidy' which made capitalism possible in the first place. It did so by enabling the creation of surpluses for investment – surpluses which, it has been estimated, might have been completely absorbed if 'environmental remediation' (repairing all damage and replacing all spent resources) had been obligatory in the past as it will be in the future.[1]

Because we have failed to be aware of the existence and nature of this 'subsidy', we have built up accounting systems to control the input/output flows across the boundary of the organization in purely money terms. The twofold task which the accounting profession is going to be called upon to tackle if the industrial economy is to keep going is a mammoth one, and the sooner it is encouraged to set about it, the better.

The first part of the task adds to accountants' workloads, but is relatively easy in principle because it simply means incorporating and integrating an additional *class* of financial flows, i.e. the quantifiable environmental investments, costs, savings and income related to operating sustainably, into the existing management and financial accounting systems.

The second part is more challenging because it will take accountants out of their traditional sphere in which all operations and changes are measured in money terms. The need to measure and balance the flows of materials and energy in and out of the organization in the same way as is done for the money flows is essentially an accounting task (as indicated by such terms as 'natural resource accounting' and 'environmental balance sheet'), calling on the same kinds of models, disciplines and attitudes of mind, and it will be a pity (and dysfunctional for business) if the accounting profession fails to pick up the challenge and a separate, possibly competing, discipline of environmental accounting has to be created.

13.6 New marketing parameters

As markets become more international and a lot more competitive, corporate strategies are becoming more and more market-driven, and are often difficult to distinguish from market strategies. It appears increasingly that environmental issues are likely to have a big and growing influence on the conditions governing international market share and penetration. Even the apparent triumph of trade liberalization over environmental protectionism in the last GATT Round was certainly not the last word on that particular subject, but rather the beginning of a long debate.

In the last five or six years marketers have begun to look to environmental product features for localized competitive advantage, and companies have tried to build up their environment-friendly image to improve their positioning in the market. Superficial observers detected a kind of 'flavour-of-the-month' effect, and were not surprised when the green consumer wave, having peaked around 1991, seemed to subside again.

The reality apparent to those willing to look more deeply is that the brief green consumer boom was only a precursor to the much more radical restructuring that has begun to take place in the spectrum of attitudes and values affecting market choice, not only by supermarket shoppers, but by business and institutional buyers under the influence of sustainability goals and changed market forces.

From a disjointed marketing 'plus', the environment is moving towards a central position of necessary integration with, or even initiation of, corporate marketing strategy. An environmental SWOT analysis can point to new business opportunities under the traditional headings of diversification, innovation and repositioning of market offerings. Other techniques, such as environmental product life-cycle (or cradle-to-grave) analysis, are becoming important marketing tools.

The outcome, within a sustainability perspective, is that marketing strategies have to become environmentally 'coherent', so that reduced product environmental impact is supported by all the other aspects of marketing, including packaging (increasingly seen as an integral product feature), promotion, distribution, brand image and price.

13.7 The special role of R and D

A strategy of sustainable business development places a special responsibility on the scientific and technical functions of R and D and new product development and design. These go well beyond the watchdog function of guarding against the serious risks to the business and its products which arise when potential environmental effects of products and new processes are not carefully and expertly monitored by qualified specialists.

This risk-avoiding function is important for every business, but not sufficient to support the more proactive strategies of 'corporate citizenship' and 'strategic opportunity', leading to a sustainable operation. These strategies, to be effective, depend on R and D and the related design and development functions taking upon themselves the goal of responding directly to environmental problems in which the company or its products may be involved, irrespective of whether they have yet become the subject of legislation or campaigning (though keeping an eye on the matters exercising green pressure groups may give useful warnings of developing issues).

Such a role shades by almost imperceptible degrees into supporting a 'strategic opportunity' approach, by extrapolating from the company's own environmental impacts and its technological strengths to opportunities for introducing new products and services which can strengthen the company's capacity to move towards sustainability.

13.8 In the front line: operations management

Although 'operations' are usually contrasted with 'strategy', the management of industrial operations constitutes a strategic input in its own right – in most industries *the* critical input as far as environmental impact is concerned. When it comes to pollution, contamination and waste, the way in which industrial installations are actually run is probably the key factor, though it is often difficult to make a clear distinction between the design and operation of a particular process.

It is possible to identify some real strategic options for operations management. One class of choices concerns the degree of proactive concern with the environmental impact of production and other operations. At one end of the scale are what is probably still the majority of operations managers, content to comply with legal requirements rather than exceed them, equating BATNEEC (Best Available Technology Not Entailing

Excessive Cost) with the ironic CATNIP (Cheapest Available Technology Not Involving Prosecution), accepting environmental regulation as a necessary evil which gets in the way of their real work.

In contrast with this minimalist approach, we can find an increasing number of management teams who, with the support of a proactive, pro-environmental company culture sponsored by top management, have become positive enthusiasts of environmental improvement, actively seeking emission reductions, energy and water savings and waste reductions as part of a strategy of reducing costs by minimizing pollution.

In such companies an active movement towards sustainable working has started at the level of operations. Other aspects of such a forward-looking strategy are often the decision to conduct regular site environmental audits, to adopt the European EMAS Regulation or the British Standard BS 7750 and to take part in inter-company benchmarking exercises and intra-industry collaborative programmes of environmental care. Such companies also encourage the active involvement of all employees in environmental improvement initiatives and support it by environmental awareness programmes of training and communication.

13.9 Extending environmental stewardship

In keeping with the outlook of ecological management, which widens the system boundaries and the stakeholder set to encompass the full range of the organization's environmental effects, the operational aspects of environmental strategy do not stop at the site boundary, but include the full life cycle of the product. In particular, this means that purchasing and logistics strategies can be important components of the overall strategy.

Most of the environmental 'leaders' have introduced, or are in the process of introducing, policies of supplier screening, auditing, and, most importantly, collaboration, with a view to minimizing their indirect environmental impacts. By tracing the origins of materials and components as far as possible along the supply chain they attempt to bring their influence as customers to bear as a means of reducing environmental impacts from earlier stages of the life cycle. By this practice of 'product stewardship', large companies who are committed to high-profile environmental strategies both avoid the embarrassment of 'guilt by association' and widen the circle of environmental care.

Similar issues arise after products leave the factory gate, in attempting to minimize the impact of transportation and related logistics operations. While the present transport infrastructures of most industrial countries do not favour the environmentally preferred option of railfreight for most loads, there is a lot that can be done in the short term to increase transport efficiency so as to reduce emission and waste impacts, and in the longer term companies can both use their influence to support the development of less damaging forms of transport and plan their operations so as to reduce the overall need for it.

13.10 Building strategic capability

In its early stages, business environmental strategy has concentrated on the concrete technical issues of reducing emissions, contamination and waste and on the regulatory and managerial issues of measurement and control systems. As a result, the human beings who, at the end of the day, have to want, and be able, to do things differently, to collaborate on finding new solutions, to move mountains of received wisdom and accepted practice, and to invent that new and unknown entity, the sustainable business, have been rather neglected.

Business values its 'human resources' and has special Human Resource Management and Development departments to secure, maintain, enhance and deploy these very special resources. Sadly, in most organizations these departments have been only minimally involved in environmental strategies or policies so far. Given the real need to mobilize people's capacities, energies and enthusiasm in aid of environmental improvement and ultimate sustainability, it seems obvious that this situation ought to change so that the special expertise of the human resource professionals in Management Development, in Training, in Personnel Management, can be brought to bear on the sustainable development project.

There is a need for their skills in training staff at all levels in general environmental awareness and in special methods for auditing, analysing and minimizing impacts, and for developing managers towards the new concepts and processes of 'ecological management' which, we believe, will be increasingly necessary. Over and above this, and the related tasks of recruitment and selection, performance appraisal, and communication and team building, there is a big job of 'change management' to be done, leading ultimately to the development of a more environmentally-friendly culture in business organizations.

13.11 Lines of communication

Because sustainability in a market economy depends on the company's commercial viability, and hence on its 'licence to operate', maintaining open lines of communication with stakeholders is of the essence. One of the basic changes that the move towards sustainability brings is that the stakeholder set to be considered is greatly enlarged. This places a special responsibility on those engaged in outward communication, in the area of public relations or 'public affairs' as companies prefer to call it these days.

Communication can never be one-way, and less than ever in the increasingly interdependent world of modern business. If sustainability demands the involvement of a wider set of stakeholders, then it is necessary to establish access to them and to find out their information needs and wishes. In order to communicate effectively it is important to become good listeners, and that means establishing situations in which two-way communication is possible. A number of ways of achieving this have been developed by market researchers and must be brought into use in corporate communications.

The further need is to find ways of validating the messages to stakeholders and confirming the legitimacy and bona fide of the communicators in these cynical times when environmental communications can only too easily be dismissed as 'greenspeak', 'greenwash' or even 'green con'. Thus communicators need to learn about the implications of such environmental techniques as environmental audit and impact assessment, product life-cycle analysis and risk assessment which can give credibility to their statements, and about the means of independent verification and endorsement available to firms.

The communications strategy which emerges must avoid the temptation to be purely defensive or reactive, warding off criticism or jumping onto fashionable bandwagons. This is a danger where the communicators are not fully aware of what the company is really doing, but for organizations which are genuine in their pursuit of sustainable operations it is not a satisfactory public relations stance and is a waste of much valuable internal information. Rather than attempting to do good by stealth, they are well advised to make use of a high-profile strategy of environmental leadership, which can help to educate their stakeholders, increase the legitimacy of sustainable development and pave the way for its realization.

13.12 Organization is the key

One of the major obstacles to implementing strategies that break radically new ground is often the absence of an *organizational* capacity to make them stick. Either no decision has been taken about who is accountable for the specific things that need to happen, with the optimistic formula 'In this company everyone is responsible for environmental protection' being more of a cop-out than a culture change. Or the functions and individuals allocated responsibility for specific aspects of environmental strategy do not have the time, the resources or the expertise needed.

The most frequent way of tackling the problem of accountability is the traditional one of appointing a person, sometimes a whole department, to take responsibility for all environmental matters. Apart from the prevailing tendency to pile the environment on top of a manager's existing brief without providing any extra time or staff, this solution fails to address the absolutely *major* requirement of any attempt to move towards a strategic approach to the environment, let alone a strategy of sustainable development.

Corporate strategy can only be decided at the level of strategic management, i.e. the top. If there is no one at board level or equivalent who is responsible for environmental strategy, then the environment will not be tackled strategically over the company as a whole, however bright and competent the titular environmental manager may be. If the board, or a senior-level coordinating group, does not take on the job of integrating the environmental strategies of the different functions or business units, then the company's approach to the environmental challenge risks being fragmented and incoherent.

In addition, there needs to be the responsibility and the clout at senior management level to frame, sanction and resource the fairly far-reaching plans that will be required to realize a company-wide proactive environmental strategy. The necessity of creating and implementing a system of environmental management through which such plans can be executed will require the presence of an environmental manager of the calibre and authority of a departmental head.

It is, of course, possible to finish up with an environmental organization which is all chiefs and no Indians. Another organizational task, therefore, is to ensure that, at the delivery end of environmental strategy, the necessary roles and responsibilities exist to ensure that the essential activities can be carried out. Broadening this concern, ways need to be found to involve staff at all levels of the company, so that top-down support for the strategy and bottom-up enthusiastic 'ownership' can meet in the middle and diffuse throughout the organization.

References and further reading

1. Rob Gray, Professor of Accounting at Dundee University and senior author of *Accounting for the Environment*, Paul Chapman Publishing, 1993, has informally voiced the opinion that if 'remediation cost' were included in the calculations Western industry would be found not to have made a profit in the last 50 years.

APPENDIX
Statements of principle

INTERNATIONAL CHAMBER OF COMMERCE: BUSINESS CHARTER FOR SUSTAINABLE DEVELOPMENT

1. Corporate Priority

To recognise environmental management as among the highest corporate priorities and as a key determinant to sustainable development; to establish policies, programmes and practices for conducting operations in an environmentally sound manner.

2. Integrated Management

To integrate these policies, programmes and practices fully into each business as an essential element of management in all its functions.

3. Process of improvement

To continue to improve corporate policies, programmes and environmental performance, taking into account technical developments, scientific understanding, consumer needs and community expectations, with legal regulations as a starting point; and to apply the same environmental criteria internationally.

4. Employee education

To educate, train and motivate employees to conduct their activities in an environmentally responsible manner.

5. Prior assessment

To assess environmental impacts before starting a new activity or project and before decommissioning a facility or leaving a site.

6. Products and Services

To develop and provide products or services that have no undue environmental impact and are safe in their intended use, that are efficient in their consumption of energy and natural resources, and that can be recycled, reused, or disposed of safely.

7. Customer Advice

To advise, and where relevant educate, customers, distributors and the public in the safe uses, transportation, storage and disposal of products provided; and to apply similar considerations to the provision of services.

8. Facilities and operations

To develop, design and operate facilities and conduct activities taking into consideration the efficient use of energy and materials, the sustainable use of renewable resources, the minimisation of adverse environmental impact and waste generation, and the safe and responsible disposal of residual wastes.

9. Research

To conduct or support research on the environmental impacts of raw materials, products, processes, emissions and waste associated with the enterprise and on the means of minimising such adverse impacts.

10. Precautionary approach

To modify the manufacture, marketing or use of products or services or the conduct of activities, consistent with scientific and technical understanding, to prevent serious or irreversible environmental degradation.

11. Contractors and suppliers

To promote the adoption of these principles by contractors acting on behalf of the enterprise, encouraging and, where appropriate, requiring improvements in their practices to make them consistent with those of the enterprise; and to encourage the wider adoption of these principles by suppliers.

12. Emergency preparedness

To develop and maintain, where significant hazards exist, emergency preparedness plans in conjunction with the emergency services, relevant authorities and the local community, recognising potential transboundary impacts.

13. Transfer of technology

To contribute to the transfer of environmentally sound technology and management methods throughout the industrial and public sectors.

14. Contributing to the common effort

To contribute to the development of public policy and to business, governmental and intergovernmental programmes and educational initiatives that will enhance environmental awareness and protection.

15. Openness to concerns

To foster openness and dialogue with employees and public, anticipating and responding to their concerns about the potential hazards and impacts of operations, products, wastes or services, including those of transboundary or global significance.

16. Compliance and reporting

To measure environmental performance; to conduct regular environmental audits and assessment of compliance with company regulations, legal requirements and these principles; and periodically to provide appropriate information to the Board of Directors, shareholders, employees, the authorities and the public.

THE CERES PRINCIPLES

(Formerly the Valdez Principles)

Introduction

By adopting these principles we publicly affirm our belief that corporations have a responsibility for the environment, and must conduct all aspects of their business as responsible stewards of the environment by operating in a manner that protects the Earth. We believe that corporations must not compromise the ability of future generations to sustain themselves.

We will update our practices constantly in the light of advances in technology and new understandings in health and environmental science. In collaboration with CERES, we will promote a dynamic process to ensure that the Principles are interpreted in a way that accommodates changing technologies and environmental realities. We intend to make consistent, measurable progress in implementing these Principles and to apply them to all aspects of our operations throughout the world.

Protection of the Biosphere

We will reduce and make continual progress toward eliminating the release of any substance that may cause environmental damage to the air, water, or the Earth or its inhabitants. We will safeguard all habitats affected by our operations and will protect open spaces and wilderness, while preserving biodiversity.

Sustainable Use of Natural Resources

We will make sustainable use of renewable natural resources, such as water, soils and forests. We will conserve non-renewable natural resources through efficient use and careful planning.

Reduction and Disposal of Waste

We will reduce and where possible eliminate waste through source reduction and recycling. All waste will be handled and disposed of through safe and responsible methods.

Energy Conservation

We will conserve energy and improve the energy efficiency of our internal operations and of the goods and services we sell. We will make every effort to use environmentally safe and sustainable energy sources.

Risk Reduction

We will strive to minimize the environmental, health and safety risks to our employees and the communities in which we operate through safe technologies, facilities, and operating procedures, and by being prepared for emergencies.

Safe Products and Services

We will reduce and where possible eliminate the use, manufacture, or sale of products and services that cause environmental damage or health and safety hazards. We will inform our customers of the environmental impacts of our products or services and will try to correct unsafe use.

Environmental Restoration

We will promptly and responsibly correct conditions we have caused that endanger health, safety, or the environment. To the extent feasible, we will redress injuries we have caused to persons or damage we have caused to the environment and will restore the environment.

Informing the Public

We will inform in a timely manner everyone who may be affected by conditions caused by our company that might endanger health, safety, or the environment. We will regularly seek advice and counsel through dialogue with persons in communities near our facilities. We will not take any action against employees for reporting dangerous incidents or conditions to management or to appropriate authorities.

Management Commitment

We will implement these Principles and sustain a process that ensures that the Board of Directors and Chief Executive Officer are fully informed about pertinent environmental issues and are fully responsible for environmental policy. In selecting our Board of Directors, we will consider environmental commitment as a factor.

Audits and Reports

We will conduct an annual self-evaluation of our progress in implementing these Principles. We will support the timely creation of generally accepted environmental audit procedures. We will annually complete the CERES Report, which will be made available to the public.

Disclaimer

These Principles establish an environmental ethic with criteria by which investors and others can assess the environmental performance of companies. Companies that sign these Principles pledge to go voluntarily beyond the requirements of the law. These Principles are not intended to create new legal liabilities, expand existing rights or obligations, waive legal defences, or otherwise affect the legal position of any signatory company and are not intended to be used against a signatory in any legal proceedings, for any purpose.

CERES 28 April 1992

Index